Women in Engineering and Science

Series Editor

Jill S. Tietjen
Greenwood Village, Colorado, USA

More information about this series at http://www.springer.com/series/15424

Stacey M DelVecchio

Editor

Women in 3D Printing

From Bones to Bridges and Everything in Between

 Springer

Editor
Stacey M DelVecchio
StaceyD Consulting, LLC
Peoria, IL, USA

ISSN 2509-6427 ISSN 2509-6435 (electronic)
Women in Engineering and Science
ISBN 978-3-030-70735-4 ISBN 978-3-030-70736-1 (eBook)
https://doi.org/10.1007/978-3-030-70736-1

This Springer imprint is published by the registered company Springer Nature Switzerland AG
The registered company address is: Gewerbestrasse 11, 6330 Cham, Switzerland

To Kerry, my partner in life, and to Carole, my mom and biggest fan.

Foreword 1

The Medici Effect and the Importance of Women in Manufacturing

In 2004, Frans Johansson published *The Medici Effect*, which emphasizes the need for diversity for innovation. His hypothesis is based on the Medici Dynasty in the fourteenth century and its role in the birth of the Renaissance. The Medici's financial support of painters, poets, scientists, architects (a variety of disciplines working closely together on diverse problems) led to an explosion of innovations. We still see this today in pockets of industries where architects work with entomologists on new designs of energy-efficient building, roboticists study nature to understand locomotion and manipulation. One of my most rewarding programs was when I worked with a microbiologist to use bacteria to synthesize magnetic nanoparticles for ferrofluids. Diversity is the fuel for innovation. Our greatest source of fuel for diversity is gender.

There are numerous trailblazers in gender diversity in sciences. Marie Curie was the first woman to win a Nobel Prize and the only person to win the Nobel Prize in two scientific fields. She was followed by her daughter, Irene Joliot-Curie, who received the Chemistry Prize in 1935. However, the statistics still show that women account for less than one third of those employed in scientific research and development across the world. But trends are changing in certain disciplines of science and engineering. When compared to other industries, including non-STEM, the representation of women among board directors in the information technology industry have experienced the sharpest increase, from 14.8% in 2018 to 17.9% in 2019 [1]. As an example, Dr. Deborah Frinke recently joined Oak Ridge National Laboratory as the director of the National Security Sciences Directorate. Dr. Fricke is a computer scientist that specialized in cybersecurity and was the director of research at NSA from 2013 to 2020. What is it about computer science that attracts women to the field? How do we catalyze gender diversity in manufacturing?

Manufacturing has always been a male-dominated profession. However, there have been exceptions. We saw during World War II that the engines of production would have halted without women in manufacturing as exemplified by Rosie the Riveter. Closer to my home institution, Oak Ridge National Laboratory (ORNL), the Manhattan Project depended upon the Calutron Girls who monitored the control panels that separated uranium isotopes. There have been trailblazers in manufacturing leadership. Dr. Susan Smyth, the 2020 President of the Society of Manufacturing Engineers (SME) is a Fellow of SME and National Academy Member. She was the chief scientist for global manufacturing at General Motors and director of their Manufacturing Systems Research Labs, directing the creation of GM's global manufacturing R&D strategies. In additive manufacturing, Dr. Dawn White invented and patented ultrasonic additive manufacturing and founded Solidica in 1999, which later became Fabrisonic. Dawn continues to be a serial entrepreneur in manufacturing. Both women are technical, innovative, and insightful, making them natural manufacturing leaders. Susan and Dawn have been manufacturing trailblazers. We now need manufacturing road pavers, more women to see the vast array of career opportunities in manufacturing. Rosie the Riveter and the Calutron Girls showed women could excel on the shop floor while Deborah, Susan, and Dawn have shown they can excel in the boardroom. Can additive manufacturing help fuel the fire for gender diversity in manufacturing?

I believe so and am optimistic. For ten years I was a mentor for the FIRST Robotics program. FIRST created an environment that encouraged diversity and hands-on collaboration where young men and women worked together solving challenging problems in a very compressed schedule (6 weeks). It truly is an intramural sport with a focus on manufacturing. For five years, Oak Ridge National Laboratory opened the Manufacturing Demonstration Facility up to these students, providing them direct access to the latest additive manufacturing tools. I saw an explosion of innovative ideas from the students when they were provided the opportunity to print parts for their robots. I remember it took just one or two confident women (younger versions of Deborah Fricke, Susan Smyth, and Dawn White) to show that they could have fun with science and engineering and they could stand shoulder to shoulder with the young men problem-solving (as well as using 3D printers and power tools). What became more interesting to me is I noticed over time that these women exhibited natural leadership skills. They were organized, creative, confident, and calm (more than me) in stressful situations. I have continued to see this today with young women I recruit and work with (one of whom is an author of a chapter in this book). Today, 29% of women, compared to 15% in 2015, believe schools actively encourage female students to pursue careers in manufacturing with 42% of women today ready to encourage their daughters or female family members to pursue manufacturing careers [2]. These are trends that excite me.

It is with this background, and my belief that diversity fuels innovation, that I am thrilled to see a book on additive manufacturing authored by women, many of whom are technical and business leaders in the additive manufacturing community.

Oak Ridge National Laboratory   Lonnie J. Love,
Oak Ridge, TN, USA

References

1. Emelianova, O., Milhomem, C.: Women on Boards: 2019 Progress Report, p. 15. (MSCI, December 2019)
2. Giffi, C., et al.: Women in Manufacturing: Stepping Up to Make an Impact That Mattters, D

Foreword 2

I first met Stacey DelVecchio during a #3DTalk panel I co-organized with Barbara Hannah, founder of Cyant, during the international conference RAPID+TCT 2018. The panel discussion was about the adoption of additive manufacturing into the overall manufacturing process. At the time, Stacey was the Additive Manufacturing Product Manager at Caterpillar and already an authority in the implementation of 3D printing in her industry. I was quite impressed by her insights and experience implementing additive manufacturing at Caterpillar.

As the founder of the Women in 3D Printing organization, I am always eager to know other female professionals in our industry, hear about their background, involvement in the additive manufacturing industry, and current contributions in our industry.

I am sincerely impressed by Stacey. From her 30-year long career at Caterpillar, managing the additive manufacturing factory, to now running her consulting firm, StaceyD Consulting, Stacey did not forget to inspire and encourage more women to follow her steps! A true advocate for women in engineering, she served as the president of the Society of Women Engineers in 2014.

Her work with this book is another testimony to her dedication for a more gender-balanced additive manufacturing industry. Each of the women you'll hear from in this book have brought a considerable contribution to the additive manufacturing industry and 3D printing technologies, whether by accelerating the use of additive manufacturing in their respective industries, pushing the boundaries of machine capacities, advancing materials, or developing key software features.

This book is about the wide-spreading adoption of additive manufacturing for industrial applications, and exploring the technical advancements that are happening to enable this wider adoption. Stacey decided to do so by featuring some of the female individuals who are contributing to these advancements. The Women in 3D Printing organization has now been around for 6 years, and I am extremely proud and thankful for the work that has been done in our industry for such a book to even exist. I am grateful to Stacey and the 10 other women featured in this book for their everyday work, facilitating the adoption of 3D printing, and sharing their work throughout this book.

Women in 3D Printing   Nora Touré,
Berkeley, CA, USA

Preface

Look online or in industry periodicals and you'll find large amounts of information about 3D printing or additive manufacturing. There's also considerable information in these same online sources about the men, women, and companies that are major contributors to the industry and technology. But there are very few published books about the contributors to this rapidly evolving industry. And there are no published books about the amazing female champions in the field. This book highlights these impressive women and their contributions to the industry of 3D printing.

My sincere thanks to all of my chapter authors who willingly volunteered to share their stories. Some of these authors I know well, some are references of colleagues, and some are simply impressive women I read about. This book would be literally nothing without them. Thanks also goes to my mentor in this process, Jill Tietjen; my sounding board and proofreader, Holly Teig; and my numerous colleagues who were always there to help with technical content or general perspective.

Peoria, IL, USA   Stacey M DelVecchio

About the Book

In this book, each chapter is written by a technical leader in 3D printing that happens to be a woman. Their stories are extremely different, not only in their industry within 3D printing, but also in the types of work they are doing. Some of the women are deep in research, some are focused on outreach, and some are managing the business of 3D printing; some of these women are just starting their careers while others are seasoned professionals. All are accomplished engineers. If you're looking to learn about the technical specifics of 3D printing, this is not the book for you. If you're looking for a glimpse into the variety that 3D printing has to offer both technically and from a career perspective, then read on to learn about what these inspiring women have done. At the end of each chapter, you will find a biography of the chapter author and co-authors.

Abbreviations

2D	Two-dimensional
3D	Three-dimensional
3DP	Three-dimensional printing or 3D printing
AAAS	American Association for the Advancement of Science
ABS	Acrylonitrile butadiene styrene
AC	Additive construction
ACES	Automated Construction of Expeditionary Structures
ACI	American Concrete Institute
ACL	Anterior cruciate ligament
ADF	Alloy development feeder
AISI	American Iron and Steel Institute
AIAA	American Institute of Aeronautics and Astronautics
Al_2O_3	Aluminum oxide
AL 6061-T6	Aluminum alloy 6061
ALP	Alkaline phosphatase
AMUG	Additive Manufacturing Users Group
AR	Augmented reality
ASCE	American Society of Civil Engineers
ASTM	American Society of Testing and Materials
AUVSI Foundation	Association for Unmanned Vehicle Systems International Foundation
AW	Apatite-wollastonite
BC	Boron carbide
BJP	Binder jet printing
C	Carbon
Ca	Calcium
CAD	Computer-aided design
CANRIMT2	Canadian Network for Research and Innovation in Machining Technology
C_f	Carbon fiber
CaP	Calcium phosphate

C$_f$/SiOC	Carbon-fiber-reinforced silicon oxycarbide
CCM	Coordinate measuring machine
CLIP	Continuous liquid interface production
CMC	Ceramic matrix composites
CNC	Computer numerical control
Co	Cobalt
CRADA	Cooperative Research and Development Agreement
CT	Computed or computerized tomography
D1, D7, D21	Day 1, day 7, day 21
D2C	Direct to customer
DASH	Direct additive-subtractive hybrid
DED	Directed energy deposition
DIW	Direct ink writing
DLP	Digital light processing
DLS™	Digital Light Synthesis™
DMD	Direct metal deposition
DMLS	Direct metal laser solidification (or sintering)
DO	Distraction osteogenesis
EB	Electron beam
EBM	Electron beam melting
EBPBF	Electron beam powder bed fusion
EPU	Elastomeric polyurethane
ERDC	Engineer Research and Development Center
ERDC-CERL	Engineer Research and Development Center, Construction Engineering Research Laboratory
FDM	Fused deposition modeling
FFF	Fused filament fabrication
FGM	Functionally gradient material
FRF	Frequency response functions
F.SWE	Fellow, Society of Women Engineers
HA	Hydroxyapatite
HAZ	Heat affected zone
HIP	Hot isostatic pressing
hTERT-MSCs	Human mesenchymal stromal cells
IJP	Ink jet printing
IP	Intellectual property
ISO	International Organization for Standardization
LENS	Laser engineered net shaping
LMD	Laser metal deposition
LPBF	Laser powder bed fusion
MC	Magnetic caloric
MCM	Magnetic caloric material
MD	Maltodextrin
MMC	Metal matrix composite
MOST	Michigan Tech Open-source Sustainability Technology

MRI	Magnetic resonance imaging
MSAM	Multi-scale additive manufacturing
MVP	Minimum viable product
NEST Tech	Next Evolution Sense Technology
NHL	National Hockey League
NRMCA	National Ready Mix Concrete Association
NSF	National Science Foundation
NSERC	Natural Sciences and Engineering Research Council of Canada
OEM	Original equipment manufacturer
ORNL	Oak Ridge National Laboratory
OSHA	Occupational Safety and Health Administration
PA	Polyamide
PAA-Na	Ammonium polyacrylate
PBF	Powder bed fusion
PCP	Preceramic polymer
PIP	Polymer infiltration and pyrolysis
PLLA	Poly-L-lactic acid
PPE	Personal protective equipment
PUF	Physical unclonable function
PVP	Polyvinylpyrrolidone
QD	Quantum dots
R&D	Research and development
RepRap	Self-REPlicating RAPid prototyper
RILEM	International Union of Laboratories and Experts in Construction Materials, Systems and Structures (from the name in French)
RPM	Revolutions per minute
TE	Tissue engineering
TEC	Thermal expansion coefficient
RM	Regenerative medicine
ROI	Region of interest
SBF	Simulated body fluid
SCC	Self-consolidating concrete
SEM	Scanning electron microscope
SIMP	Solid Isotropic Material with Penalty
Si_3N_4	Silicon nitride
SiC	Silicon carbide
SiC/SiC	Silicon carbide fiber in silicon carbide matrix
SiC_f	SiC fiber
SiO_2	Silicon dioxide
SiOC	Silicon oxycarbide
SLA	Stereolithography
SLM	Selective laser melting
SLS	Selective laser sintering

SM	Subtractive manufacturing
SMA	Shape-memory alloy
SNR	Signal to noise
SNS	Spallation neutron source
SOM	Skidmore, Owings, and Merrill
SS	Stainless steel
STL	File format for 3D printing
STEM	Science, technology, engineering, and mathematics
STEAM	Science, technology, engineering, arts and mathematics
SWE	Society of Women Engineers
Ti	Titanium
TiC	Titanium carbide
Ti-6Al-4 V (Ti64)	A titanium alloy
TRB	Transportation Research Board
UHTC	Ultrahigh-temperature ceramic
UHTCMC	Ultrahigh-temperature ceramic matrix composite
USACE	US Army Corps of Engineers
UV	Ultraviolet
UV-VIS	Ultraviolet-visible spectroscopy
VR	Virtual reality
WC	Tungsten carbide
WC-Co	Tungsten carbide infiltrated with cobalt
Wi3DP	Women in 3D Printing
XCT	X-ray computed tomography

Contents

Contributors

Amy Alexander Mayo Clinic, Department of Radiology, Anatomic Modeling Laboratory, Rochester, MN, USA

Erin Winick Anthony Sci Chic, Houston, TX, USA

Erika Berg Carbon, Inc., Redwood City, CA, USA

Stacey M DelVecchio StaceyD Consulting, LLC, Peoria, IL, USA

Amy Elliott Oak Ridge National Laboratory, Knoxville, TN, USA

Kaan Erkorkmaz University of Waterloo, Waterloo, ON, Canada

Mark Kirby Renishaw, Waterloo, ON, Canada

Megan A. Kreiger U.S. Army Engineer Research and Development Center, Champaign, IL, USA

Melanie A. Lang FormAlloy Technologies, Inc., Spring Valley, CA, USA

Priscila Melo Politecnico di Torino, Torino, Italy

Ahmet Okyay University of Waterloo, Waterloo, ON, Canada

Lisa Rueschhoff Air Force Research Laboratory, Dayton, OH, USA

Mihaela Vlasea University of Waterloo, Waterloo, ON, Canada

Yanli Zhu University of Waterloo, Waterloo, ON, Canada

Chapter 1
Introduction

Stacey M DelVecchio

Introduction

Emerging technologies are typically subject to public hype and 3D printing (i.e. additive manufacturing) is no exception. At times, people said 3D printing was going to take over the world. We were going to be printing everything, everywhere. At the same time, others were saying the technology would never get out of the prototype phase. Of course, the reality lies somewhere in between and many a heated debate has been had on what that reality actually is. For those of us working in the world of 3D printing, we have all encountered people that feel passionate about both extremes of the present state and the future of the industry.

But let's step back and level set on what we mean by 3D printing versus additive manufacturing. You'll see the chapter authors of this book use both terms interchangeably which is common in the industry. The terms "additive manufacturing" and "3D printing" both refer to creating an object by sequentially adding build material in successive cross-sections, one stacked upon another [1]. Additive manufacturing is where you add material versus subtractive manufacturing (i.e., machining), where you subtract material. Additive manufacturing tends to be the more inclusive term and is generally associated with industrial applications. From an engineering specification perspective, the American Society of Testing and Materials (ASTM) and the International Organization for Standardization (ISO) refer to the seven families of additive manufacturing seen in Figs. 1.1 and 1.2 as the broader descriptor, rather than the seven families of 3D printing. As additive manufacturing gained the attention of the media when some of the early desktop printers hit the market in the early 1990s, the term "3D printing" became popular [2]. The 3D printing terminology was compared to a 2D inkjet printer connected to a computer, making the term "3D printing" easier to visualize. This is why you will often see the

S. M DelVecchio (✉)
StaceyD Consulting, LLC, Peoria, IL, USA
e-mail: staceyd.in.3dp@gmail.com

© Springer Nature Switzerland AG 2021   1
S. M. DelVecchio (ed.), *Women in 3D Printing*, Women in Engineering and
Science, https://doi.org/10.1007/978-3-030-70736-1_1

7 Families of Additive Manufacturing
According to ISO/ASTM52900-15 (formerly ASTM F2792)

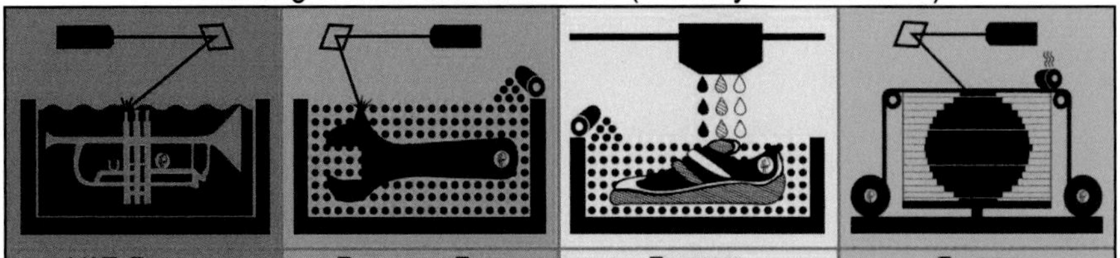

VAT PHOTO-POLYMERIZATION	POWDER BED FUSION (PBF)	BINDER JETTING	SHEET LAMINATION
Alternative Names SLA™- Stereolithography Apparatus DLP™- Digital Light Processing 3SP™- Scan, Spin, and Selectively Photocure CLIP™- Continuous Liquid Interface Production	**Alternative Names** SLS™- Selective Laser Sintering DMLS™- Direct Metal Laser Sintering SLM™- Selective Laser Melting EBM™- Electron Beam Melting SHS™- Selective Heat Sintering MJF™- Multi-Jet Fusion	**Alternative Names** 3DP™- 3D Printing ExOne Voxeljet	**Alternative Names** Polyjet™ SCP™- Smooth Curvatures Printing MJM - Multi-Jet Modeling Projet™
Description A vat of liquid photopolymer resin is cured through selective exposure to light (via a laser or projector) which then initiates polymerization and converts the exposed areas to a solid part.	**Description** Powdered materials is selectively consolidated by melting it together using a heat source such as a laser or electron beam. The powder surrounding the consolidated part acts as support material for overhanging features.	**Description** Liquid bonding agents are selectively applied onto thin layers of powdered material to build up parts layer by layer. The binders include organic and inorganic materials. Metal or ceramic powdered parts are typically fired in a furnace after they are printed.	**Description** Droplets of material are deposited layer by layer to make parts. Common varieties include jetting a photcurable resin and curing it with UV light, as well as jetting thermally molten materials that then solidify in ambient temperatures.
Strengths • High level of accuracy and complexity • Smooth surface finish • Accommodates large build areas	**Strengths** • High level of complexity • Powder acts as support material • Wide range of materials	**Strengths** • Allows for full color printing • High productivity • Uses a wide range of materials	**Strengths** • High level of accuracy • Allows for full color parts • Enables multiple materials in a single part
Typical Materials UV-Curable Photopolymer Resins	**Typical Materials** Plastics, Metal and Ceramic Powders, and Sand	**Typical Materials** Powdered Plastic, Metal, Ceramics, Glass, and Sand	**Typical Materials** Paper, Plastic Sheets, and Metal Foils/Tapes

Fig. 1.1 7 Families of Additive Manufacturing (Image 1 of 2) – Four of the seven families of additive manufacturing, per ASTM F2792 [3]

7 Families of Additive Manufacturing - continued
According to ISO/ASTM52900-15 (formerly ASTM F2792)

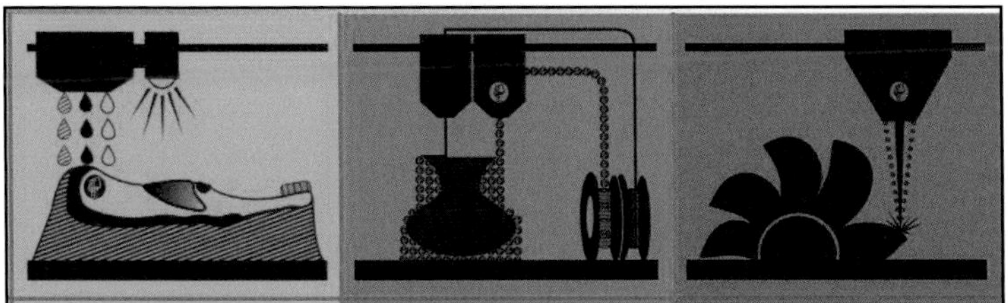

MATERIAL JETTING	MATERIAL EXTRUSION	DIRECTED ENERGY DEPOSITION (DED)
Alternative Names Polyjet™ SCP™- Smooth Curvatures Printing MJM - Multi-Jet Modeling Projet™	**Alternative Names** FFF - Fused Filament Fabrication FDM™- Fused Deposition Modeling APD™ - Augmented Polymer Deposition	**Alternative Names** LMD - Laser Metal Deposition LENS™- Laser Engineered Net Shaping DMD - Direct Metal Deposition Laser cladding
Description Droplets of material are deposited layer by layer to make parts. Common varieties include jetting a photocurable resin and curing it with UV light, as well as jetting thermally molten materials that then solidify in ambient temperatures.	**Description** Material is extruded through a nozzle or orifice in tracks or beads, which are then combined into multi-layer models. Common varieties include heated thermoplastic extrusion (similar to a hot glue gun) and syringe dispensing.	**Description** Powder or wire is fed into a melt pool which has been generated on the surface of the part where it adheres to the underlying part or layers by using an energy source such as a laser or electron beam. This is essentially a form of automated build-up welding.
Strengths • High level of accuracy • Allows for full color parts • Enables multiple materials in a single part	**Strengths** • Inexpensive and economical • Allows for multiple colors • Can be used in an office environment • Parts have good structural properties	**Strengths** • Not limited by direction or axis • Effective for repairs and adding features • Multiple materials in a single part • Highest single-point deposition rates
Typical Materials Photopolymers, Polymers, Waxes	**Typical Materials** Thermoplastic Filaments and Pellets (FFF); Liquids, and Slurries (Syringe Types)	**Typical Materials** Metal Wire and Powder, with Ceramics

Fig. 1.2 7 Families of Additive Maufacturing (Image 2 of 2) – Three of the seven families of additive manufacturing, per ASTM F2792, plus hybrid additive manufacturing [3]

term "3D printing" used for the general public while additive manufacturing is used in industrial settings. One thing is certain though: as the industry continues to evolve, there will probably be another term applied to the technology, so for now, we'll stick with using the terms "3D printing" and "additive manufacturing."

Whether you call it 3D printing or additive manufacturing, the benefits of the technology are widespread. Additive manufacturing can be essential in the prototype phase, providing the ability to have quick design iterations using 3D printed parts. In the design phase, the technology can provide a high degree of customization as it's possible to print just a few pieces instead of what would traditionally be large lot sizes. We can design features with engineering benefits that were previously impossible to manufacture. And we can even use additive for production. Each chapter author has taken advantage of more than one of these benefits which is part of what makes the technology so interesting.

As you read through the accomplishments of each of the chapter authors and see how they are applying the technology today, it's easy to see why there continues to be hype around 3D printing. There's still a lot of work to be done and subsequent inventions that need to happen. But I'm positive, as we look to the future, the machines will be more reliable, the raw materials will be more available, the printing processes will be more repeatable and faster, and oh yes, all of this will become more affordable. Is this a short-term future? No. But it is an exciting and viable future for this technology.

References

1. GE Additive "Additive 101. http://www.ge.com/additive. Accessed 09 Dec 2020
2. SME. http://www.sme.org. Accessed 09 Dec 2020
3. Hybrid Manufacturing Technologies "7 Families of Additive Manufacturing". http://www.hybridmanutech.com. Accessed 09 Dec 2020

Stacey M DelVecchio is the president of Stacey D Consulting, focused on the business of additive manufacturing. She is a technical advisor with the Society of Manufacturing's Additive Manufacturing Community and has been an industry peer review panelist for Oakridge National Labs Manufacturing Development Facility. DelVecchio is a sought-after speaker on the value of additive and has been quoted in numerous technical magazines on 3D printing. Prior to launching her own consulting business, she was the additive manufacturing product manager for Caterpillar Inc. where her team leveraged the technology in all spaces, including new product introduction, supply chain, and operations. Her team focused on deploying the technology by working with Caterpillar product groups on design for additive. DelVecchio also managed Caterpillar's Additive Manufacturing Factory.

In her 30 years in industry, most of which was at Caterpillar, DelVecchio held numerous positions in engineering and manufacturing, including process and product development for nonmetallic components, production support for paint and process fluids, and the build and start-up for a green field facility in China. She is a certified 6 Sigma Black Belt, earning that classification by working on projects that included lean manufacturing, failure analysis, and employee engagement. Prior to her role in additive manufacturing, DelVecchio was the hose & coupling engineering manager with global responsibility for design, as well as the new product introduction manager for Cat Fuel Systems. She managed the project management office for Caterpillar engines with responsibility for the project management of all new product introduction programs, continuous improvement projects, and cost reduction project in the division. DelVecchio also developed an engineering pipeline strategy to ensure the best engineering talent was available to meet global enterprise needs.

DelVecchio holds a B.S. in chemical engineering from the University of Cincinnati in Ohio, USA. She was the 2013–2014 president of the Society of Women Engineers. She was named a fellow of the society in 2015 and received the society's Advocating Women in Engineering Award in 2018. DelVecchio was a vice-chair of the Women in Engineering Committee for the World Federation of Engineering Organization, representing the Engineering Societies of America and continues to champion global expansion of the Society of Women Engineers. She has spoken in a dozen countries on the value of gender diversity in engineering and is a strong advocate for women in engineering. In 2019, she published her first e-book, entitled *I Wish…*. It is a quick read of a collection of her wishes for the engineering community based on her 30+ years as an engineering leader that happens to be a woman.

Chapter 2
Medical 3D Printing: Patient-Specific Anatomic Models

Amy Alexander

Introduction

3D printing has applications in medicine that reach from anatomical and biological education to clinical regenerative medicine. Artists, illustrators, designers, engineers, physicians, and scientists have acknowledged and leveraged the use of additive machines for decades, and rely heavily on the technology to move from concept to prototype to final part in today's healthcare ecosystem. Advanced image processing allows experts to depend on radiological data to accurately represent and recreate the internal components of the body. As a senior biomedical engineer in the Mayo Clinic Anatomic Modeling Laboratory, I have the opportunity to build anatomic models from radiological data to improve understanding and surgical planning. This chapter discusses the variety of applications 3D printing has at the point of care and anatomic modeling from my perspective.

Radiology

Humankind has endeavored to map the structures within the body since, or perhaps even prior to, the advent of hieroglyphics [1]. A certain fascination surrounding what the body is made of and how the body functions has caught the attention of the world's early anatomists and physicians since the Renaissance [2]. In the sixteenth century, the use of hand-sculpted wax figures, which simulated both healthy anatomy and trauma cases for training medical responders, was adopted [3]. A significant development in viewing the body internally without surgical exploration is the

A. Alexander (✉)
Mayo Clinic, Department of Radiology, Anatomic Modeling Laboratory,
Rochester, MN, USA
e-mail: Alexander.Amy@mayo.edu

© Springer Nature Switzerland AG 2021   7
S. M. DelVecchio (ed.), *Women in 3D Printing*, Women in Engineering and
Science, https://doi.org/10.1007/978-3-030-70736-1_2

development of modern-day medical imaging modalities, which are the methods by which internal photographs of the body are acquired. Examples of these imaging modalities are computed tomography (CT) and magnetic resonance imaging (MRI) scanning, which create internal pictures of the body in a volumetric fashion. These modalities have revolutionized the methods by which anatomy can be rendered in three dimensions.

Imaging Modalities and Acquisition

After the invention of the X-ray in 1895, the subsequent developments of CT scanning, MRI scanning, and volumetric ultrasound data acquisition have allowed for exceedingly accurate and 1:1 scale renderings of patient-specific anatomy [4]. When seeking a digital reconstruction with the highest detail, a CT or MRI scan with thin slices and a high signal-to-noise (SNR) ratio is preferred [5]. A single X-ray cannot be used to create a 3D reconstruction as it is only one slice and there is no second slice with which to interpolate a 3D mesh. A volumetric ultrasound scan, also known as a 3D ultrasound, may be useful to generate several facets of a 3D structure, but it is unlikely to capture all facets in 360 degrees due to the sonic nature of data acquisition. Figure 2.1 shows how the pixels within a CT scan of the head can be isolated from the remaining data and then interpolated in all three dimensions to create a 3D mesh of the skull.

Image Segmentation

In order to visualize any internal structure that was captured in medical imaging, a step called image segmentation must be completed. Segmentation is the act of isolating the pixels within the imaging that represent a specific part or region-of-interest (ROI); this ROI is sometimes called a mask. Since CTs and MRIs are displayed as two-dimensional (2D) arrays slice by slice, it is necessary to use the tools of segmentation to capture the pixels that comprise each anatomic structure through all slices into masks. Then, once all pixels that represent a desired ROI are encapsulated together in a mask, image processing programs are used to interpolate between the selected pixels and generate a surface in three dimensions (Fig. 2.2).

For example, in order to digitally render the shape and extent of a patient's airway into their lungs, the pixels that represent air must be collected into a single subset, ROI. In a CT scan, air is the least dense substance captured in the imaging, so it will inherently be assigned the lowest, or least dense, value in the scan. To capture air, one may use a common segmentation tool called thresholding, and keep the maximum value sub-soft tissue, as shown in Fig. 2.3.

As needed, other tools like automatic pixel expansion or seed-planting isolation may be used to ensure the accuracy of the pixel subset. In more advanced software,

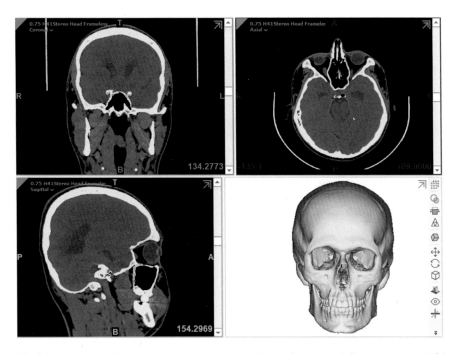

Fig. 2.1 Illustration of skull segmentation and interpolation of segmented slices to create an ROI and resultant 3D mesh of an adult skull using a CT scan. Segmentation of the imaging performed by former Mayo Clinic undergraduate researcher, Ms. Anna Sofia Stans, 2017

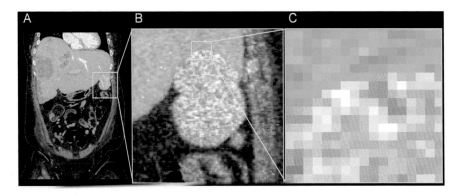

Fig. 2.2 Single CT scan slice and segmentation of an adult liver. (**a**) a view of the same sliced zoomed it to focus on where the liver meets the left kidney; (**b**) a view of the same sliced zoomed in further; and (**c**) a final zoomed view to show the pixels selected in the liver segmentation. Segmentation of the imaging performed by former Mayo Clinic undergraduate researcher, Ms. Amika Kamath, 2015

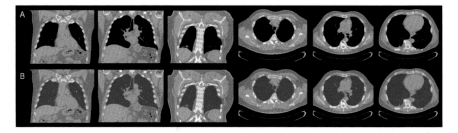

Fig. 2.3 Isolation of all bilateral lung pixels into one subset using a Hounsfield unit (HU) thresholding tool. (**a**) Un-segmented imaging data; (**b**) lungs segmented from said imaging data. Segmentation of the imaging performed by former Mayo Clinic undergraduate researcher, Ms. Claire McLellan-Cassivi, 2018

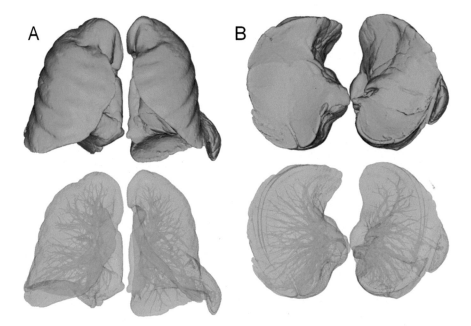

Fig. 2.4 Adult bilateral lung mesh calculated from image segmentation in Fig. 2.3. (**a**) Anterior (front-facing) view of the lungs; (**b**) axial (bottom axis, also known as worm's-eye) view of the lungs. Segmentation of the imaging and calculation of 3D mesh performed by former Mayo Clinic undergraduate researcher, Ms. Claire McLellan-Cassivi, 2018

automatic segmentation may shorten the user time, but hands-on work is still required in most cases to create accurate models. Once complete, the set of pixels on each slice is interpolated to generate a 3D surface by the use of a calculation algorithm. The final result of this segmentation is a 3D part, or model, which represents the patient's airway (Fig. 2.4).

Mesh Creation and Optimization

Depending on the use of the renderings, additional structures can be isolated in segmentation and 3D parts can be made. In some cases, multiple anatomies are of interest; when the goal is to print an anatomic model with multiple structures, each one must be segmented and each 3D part must be generated. In many instances, the part generated through automatic calculation of an input ROI is not immediately suitable for 3D printing. The first step to optimizing a mesh is to diagnose any problems that may lead to a print failure. To do so, the part generated must be exported, typically as an STL file, and imported to a computer-aided-design (CAD) software. Common culprits are inverted normals, overlapping triangles, intersecting triangles, floating shells or "floaters", or open contours. If contiguous and without problematic triangular overlaps or intersections of the polygons, this surface can be exported as an STL or other file and 3D printed.

To illustrate this point, a 3D mesh is calculated from the left femur mask in Fig. 2.5. This process incorporates an interpolation algorithm to form a surface between all selected pixels in all three dimensions. Once a mesh is generated, it can be imported into a CAD environment for further fine-tuning before preparing for 3D print.

The triangulation of the mesh must be considered healthy before it can be accepted by a 3D printer. Figure 2.6 shows a selection of marked triangles on the femoral head as well as the outcome of the mesh should those triangles be removed. When a mesh is left with an open contour, as shown in Fig. 2.6a, the mesh is no longer a water-tight surface and will not be accepted by the 3D printer's slicing software. Other common problems with triangulation at this stage are triangles that overlap or intersect one another. Additionally, triangles with internal surfaces facing

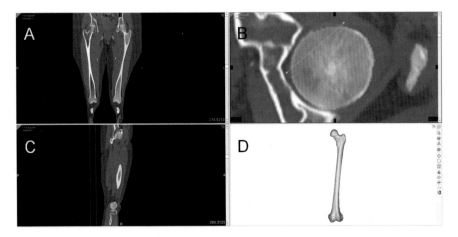

Fig. 2.5 Segmentation process from CT scan to ROI to 3D mesh of a left femur. (**a**) Coronal, (**b**) axial, (**c**) sagittal, and (**d**) 3D view of segmented left femur mesh

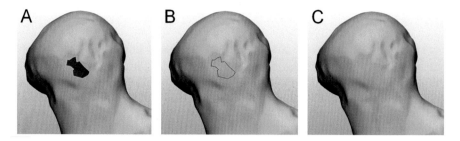

Fig. 2.6 Reviewing the triangular mesh and determining overall mesh health. (**a**) Missing triangles on the surface of the femoral head mesh, (**b**) hole filled to create a new surface and complete the mesh, and (**c**) surfaces merged to reveal a final watertight part

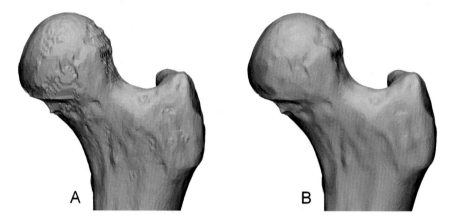

Fig. 2.7 Smoothing a mesh using CAD. (**a**) Original segmented left femur mesh. (**b**) Smoothed left femur mesh

outward, where the normal is inverted, must be corrected. Such mesh issues must be corrected.

Additional tools can be used to optimize the mesh and decrease the appearance of artifact due to the medical imaging slice thickness or noise. Figure 2.7 shows the effects of a smoothing algorithm on the femoral head mesh.

Stamping the model with a unique identifier is crucial to traceability. Many CAD applications have several options, including an embossing or embedding tool as shown below (Fig. 2.8).

Before sending the mesh file to the printer, the triangulated mesh should be checked to ensure there are no holes, marked triangles, inverted, or intersecting triangles. In addition, contours of the final version should be reviewed over the imaging by a radiologist. This means the file needs to be exported from the CAD environment and imported back into the segmentation environment for review. Note that some applications contain both segmentation and CAD in one, which simplifies this process. The contour review at this stage is recommended to ensure that the

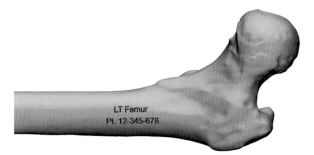

Fig. 2.8 Stamped identifier to trace this model back to the patient and/or imaging series used and laterality of the femur bone (left)

smoothing of the mesh along with fixing of the triangles to ensure a healthy mesh did not alter the file in a way that would be inconsistent with the patient's anatomy. Once identified and corrected, a water-tight part can be moved to the second step of optimization, which is typically a smoothing or wrapping function to average out surface noise.

Applications: Digital and Physical

Once the anatomic structures are prepared, they can be utilized in a variety of ways. The 3D parts can be brought up on a 2D monitor and rotated in a faux 3D space to show a patient or medical team. The 3D parts can be brought into a virtual reality (VR) space where the inner nooks and crannies of the anatomy can be explored. The 3D parts can be brought into an augmented reality (AR) space where they can be overlaid upon a patient in clinic or on the operating table. Finally, the 3D parts can be 3D printed to provide a life-size (or scaled) model of the anatomy with a tactile component. To bring an additional level of authenticity for educational purposes, surgical specimens that have been resected, or removed, from a patient can be surface scanned using photogrammetry or structured light scanning, and the resulting geometries and colors captured can be displayed in augmented reality, like the heart specimen shown in Fig. 2.9.

With new developments in biomimicry from 3D printing materials scientists, printed anatomic models comprise patient-specific "tissues" with lifelike mechanical properties. The combination of advanced medical imaging, segmentation and CAD software, and 3D printing technology, the field of anatomy education continues to advance.

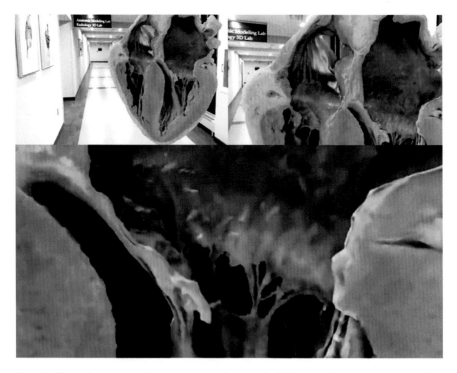

Fig. 2.9 3D surfaced scanned heart specimen displayed in AR in the hallways of the Mayo Clinic Anatomic Modeling and Radiology 3D Labs, Dr. Jonathan Morris, Medical Director, Anatomic Modeling Laboratory, Mayo Clinic, 2020

Surgery

The current and future surgical applications of 3D printing are vast. 3D printed anatomic models from radiological data for visual aid can help with patient comprehension and improved consent, as well as facilitate deeper conversations and planning among surgical teams prior to entering the operating room (OR). When prepared with materials that have similar mechanical properties to different tissues, anatomic models are also used for high fidelity surgical rehearsal or task training for surgical residents and fellows. 3D printed patient-specific guides and implants of polymers, metals, or bioprinted materials add an element of personalized medicine that has only been possible since the late 1980s. Outside of patient-specific 3D printed models, guides, and implants is the field of biomedical device design and manufacture, which 3D printing has revolutionized over the last 3 decades. The section of the chapter will discuss these applications illustrated through clinical examples.

Patient Comprehension with Anatomic Models

While a black and white plain film X-ray or volumetric scans like computed tomography or magnetic resonance imaging on a flat panel monitor are helpful to medical professionals who know how to read them, they are not always clear to patients or family members during a clinical consult. Figure 2.10 shows how a basic image segmentation and conversion of medical imaging to 3D mesh and then 3D print can assist with basic comprehension of the extent of a tumor.

Physician Surgical Planning with Anatomic Models

As mentioned previously, the 3D mesh can be used in virtual or augmented reality platforms to view and learn about the anatomic structures and discuss surgical approach options. That said, there are two main reasons that surgeons prefer a 3D printed model to an on-screen representation:

1. *Scale*: When objects are shown on-screen, the scale of the elements are often either scaled up or scaled down for better visualization; but a life-sized 3D printed anatomic model instantly relays the size of both the defect and the surrounding tissues.
2. *Tactility*: There is an innate and powerful connection between hands, eyes, and brain, so having a physical object to hold and review is a helpful feature for surgeons.

Both the scale and tactility elements can be shown in Fig. 2.11, which shows a US quarter coin next to a 3D printed model of the blood vessels surrounding a pediatric heart as well as the young patient's airway branches.

A 3D printed anatomic model allows the lead surgeon to formulate a plan with colleagues and then quickly and concisely convey that plan to the entire surgical

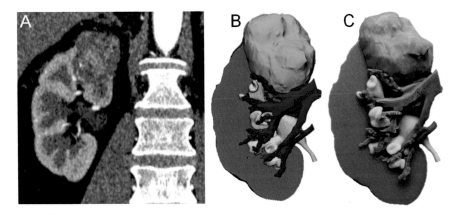

Fig. 2.10 (**a**) CT scan, (**b**) 3D mesh rendering, and (**c**) 3D print of kidney tumor

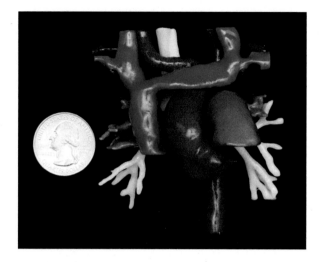

Fig. 2.11 Pediatric airway and chest vessel model next to a US quarter coin to show scale while the 3D printed model has inherent tactility

team, which may include medical doctors (MDs) like surgeons or anesthesiologists, certified registered nurse anesthetists (CRNAs), registered nurses (RNs), certified surgical technicians (CSTs), physician's assistants (PAs), nurse practitioners (NPs), and residents or fellows who are learning from the lead MDs.

Surgical Simulation and Task Training

Beyond a simple visual aid anatomic model made with 3D printing is the idea of an anatomic model whose parts have mechanical properties that are similar to biological tissues. One way to achieve a material that is bone-like is to print in a gypsum powder feedstock on a binder jetting printer and infiltrate the resulting prints with paraffin wax. The resulting model, according to orthopedic surgeons at Mayo Clinic, feels and behaves similar to bone when cut into with a bone saw or drilled into to place pins and screws commonly used in orthopedic reconstruction surgeries. For example, an adult left elbow which has previously undergone orthopedic surgery, including a humerus bone plate and an elbow joint hinge implant, is segmented from a CT scan and converted into a 3D printed elbow model that allows for extension and flexion that matches the patient's current joint capability. The orthopedic surgeon can then examine the current state of the elbow and make osteotomies, or bone cuts, to reconstruct the elbow on the model in a surgical rehearsal (Fig. 2.12).

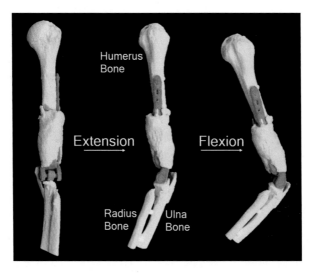

Fig. 2.12 Adult left elbow joint (tan) with previous titanium implants (blue) 3D printed on a binder jetting printer (ProJet 660Pro, 3D Systems, Rock Hill, SC, USA) and infiltrated in paraffin wax to allow for a bone-like feel and behavior during simulated orthopedic procedures; one elbow model shown in three different states of musculoskeletal extension and flexion

Additional Applications

The ability to build intricate parts with high degrees of complexity makes 3D printing a feasible option for manufacturing some of the most advanced devices. Medical device manufacturing is adopting 3D printing quickly as engineers recognize the capabilities technologies, like powder bed fusion, offer in terms of surfacing and lattice structure creation [6]. Institutions across the world are also using 3D printing to build or augment the generation and survival of bio-tissues in scaffolds; some trials have even reached the level of animal testing for long-term human viability analysis [7].

References

1. The art of healing in ancient Egypt: a scientific reappraisal – ScienceDirect. Accessed December 6, 2020. https://www.sciencedirect.com/science/article/pii/S0140673608617493?via%3Dihub
2. The evolution of anatomical illustration and wax modelling in Italy from the 16th to early 19th centuries – Riva – 2010 – Journal of Anatomy – Wiley Online Library. Accessed December 6, 2020. https://onlinelibrary.wiley.com/doi/full/10.1111/j.1469-7580.2009.01157.x
3. Mortimer Frank, Johann Ludwig Choulant, and the history of anatomical illustration - Robert M Feibel, 2019. Accessed December 6, 2020. https://journals.sagepub.com/doi/10.1177/0967772017708648
4. From Röntgen to Magnetic Resonance Imaging | North Carolina Medical Journal. Accessed December 6, 2020. https://www.ncmedicaljournal.com/content/75/2/111
5. Radiological Society of North America (RSNA) 3D printing Special Interest Group (SIG): guidelines for medical 3D printing and appropriateness for clinical scenarios | 3D Printing in Medicine | Full Text. Accessed December 6, 2020. https://threedmedprint.biomedcentral.com/articles/10.1186/s41205-018-0030-y

6. 3D-printing techniques in a medical setting: a systematic literature review | BioMedical Engineering OnLine | Full Text. Accessed December 20, 2020. https://biomedical-engineering-online.biomedcentral.com/articles/10.1186/s12938-016-0236-4
7. Frontiers | 3D Bioprinting and the Future of Surgery | Surgery. Accessed December 20, 2020. https://www.frontiersin.org/articles/10.3389/fsurg.2020.609836/full

Amy Alexander is a senior biomedical engineer in the Department of Radiology's Anatomic Modeling Unit at Mayo Clinic in Rochester, Minnesota, USA. In her role, Amy uses advanced medical segmentation and computer-aided design software to convert 2D radiological imaging data into 3D printed anatomic models. The 3D models form a communication bridge for patients and families. She also works directly with surgeons to digitally plan oncologic and orthopedic reconstructive surgery. Once digital plans are finalized by the surgeons, Amy designs and manufactures patient-specific surgical anatomic guides that are 3D printed and sterilized, then used in the operating room to help carry out the surgical plan. These patient-specific anatomic models and guides help surgeons prepare for, practice, and perform complex procedures.

Amy chairs the Engineering Education subcommittee of the Radiological Society of North America's 3D Printing in Medicine Special Interest Group (RSNA 3D SIG). She is an advisor on the American Society of Mechanical Engineering's (ASME) Medical AM committee. She is chair of the SME Medical Additive Manufacturing3D Printing Workgroup. Amy holds a bachelor of science degree in biomedical engineering from the Milwaukee School of Engineering (MSOE), and a master of science degree in engineering management from MSOE's Rader School of Business. She holds a certificate in MIT's Additive Manufacturing for Innovative Design and Production as well as an Additive Manufacturing Certification from SME. Amy is an active manuscript reviewer for Springer Nature's 3D Printing in Medicine Journal. She has given national and international keynote speeches on medical 3D printing at the point-of-care, as well as over 60 didactic talks at international conferences, colleges, STEM achievement ceremonies, community outreach programs, and medical and additive manufacturing trade shows. Amy is a contributing co-author on more than 15 medical 3D printing research papers, technical notes, and case reports. She serves on the MSOE Biomedical Engineering Industrial Advisory Committee.

In 2019, Amy was recognized as one of 14 international recipients of the SME Outstanding Young Manufacturing Engineer Award. In 2020, she had the pleasure of being featured as a SciGirl with PBS Twin Cities Public Television. She is fortunate to have been trained and mentored by industry giants: Dr. Jane Matsumoto, Dr. Jonathan Morris, Lauralyn McDaniel, Dr. Nicole Wake, Dr. Frank Rybicki, Dr. Justin Ryan, Todd Pietila, and Andy Christensen. Amy also comes from a long line of engineers, inventors, and strong women.

Amy has had the privilege of guiding and mentoring a great number of students at the high school, undergraduate, and graduate education level. Her true passion is helping young people find their purpose. Amy is honored to contribute to this publication highlighting the work of women in the world of 3D printing.

Chapter 3
Designing for Performance and Protection with Digital Manufacturing

Erika Berg

Introduction

Polymer 3D printing is widely known for its contributions to the prototyping and low volume production fields. The ability to quickly design, manufacture, and iterate enables designers, engineers, and hobbyists alike to test concepts and make low volumes of products at relatively affordable costs. With common limitations, most polymer 3D printers (at best) complement injection molding in the prototyping and pre-scale phases.

But what if a polymeric 3D printer supported a grander contribution to Industry 4.0? What if an additive manufacturing platform created millions of finished parts, simplified final assembly, and kept an end-to-end quality control digital thread? With access to infinite data in the cloud, how would designers and engineers balance seemingly endless iterations with the practical limitations of physical testing, user feedback, and opportunity cost?

As the Managing Director of Business Development at Carbon, I've had the pleasure of leading customers and partners through these questions while they implement the Carbon platform into their designs and operations.

Carbon's product suite is so much more than a one-size fits all printer – it's a true digital product development platform that helps customers bring better products to market in less time. At Carbon, teams of engineers, scientists, and manufacturing experts create the hardware, materials, and software to provide our customers with a platform that enables parts to be designed *and* manufactured as they're meant to be.

Carbon Digital Light Synthesis (Carbon DLS™) is a patented additive manufacturing technology that uses ultraviolet light, oxygen permeable optics, and programmable liquid resins to produce parts with tunable mechanical properties and production quality surface finishes (Fig. 3.1).

E. Berg (✉)
Carbon, Inc., Redwood City, CA, USA
e-mail: erika.in.3dp@gmail.com

© Springer Nature Switzerland AG 2021
S. M. DelVecchio (ed.), *Women in 3D Printing*, Women in Engineering and
Science, https://doi.org/10.1007/978-3-030-70736-1_3

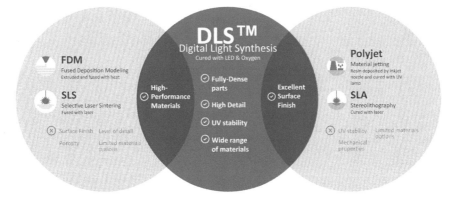

Fig. 3.1 Carbon DLS™ is a proprietary technology that fuses light and oxygen to rapidly produce products from a pool of resin

With the Carbon Design Engine™, customers can create intricate and unique lattices with structures that could not be designed by hand or using traditional CAD tools. Using the elastomeric polyurethane (EPU) materials, simple lattices printed with DLS™ have foam-like properties; but advanced structures with variable density sections, gradient transitions, and surface textures feel like custom super-foams.

My team has partnered with bleeding-edge brand power-houses like Riddell, CCM, and Specialized to use digital manufacturing at unprecedented speed and scale, enabling them to break into new categories of performance. Products incorporating DLS™ have resulted in improved comfort, protection, and reduced weight – designed with some of the coolest aesthetic features consumers have seen.

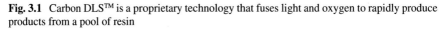

"Simple lattices printed with DLS have foam-like properties; but advanced structures with variable density sections, gradient transitions, and skin or surface textures feel like custom super-foams."

I'm thrilled by the opportunity to share how we engage with innovative partners to create sell-out products. I'll highlight one of our recent launch successes and provide insights on lessons learned along the way. Ultimately, I hope that my experiences encourage fellow female engineers, designers, and leaders in additive manufacturing to confidently push boundaries and further contributions to this field.

Professional Background

While my career in 3D printing began just a few years ago, my experiences with additive manufacturing span throughout my career, back to my days in the classroom. As an aerospace engineering student, I created scaled prototypes of early design concepts that I dreamed would once fly. As an engineer designing single-aisle commercial aircraft for The Boeing Company, I partnered with designers on the engine team that used metal printing for intricate parts within the turbofans. As a management consultant in the supply chain strategy practice at Deloitte, my most memorable engagement was with a global 3D printing company, where I led an effort to prioritize the highest value product development opportunities that would enable them to improve their use cases, gain market share, and grow revenue.

I feel fortunate to have found a role that pulls from the culmination of my past experiences. When I first joined Carbon, my initial role was a program manager; I focused on potential production application opportunities while leading customers from initial concept, through the development process, to market launch.

As a program manager, I loved partnering with innovative OEM designers to bring breakthrough ideas to life, while working with cross-functional technical and business leaders to understand customer requirements, set program goals, define critical technical specifications, and ultimately launch great products. In 2018, I led the engagement with Riddell to launch the first 3D printed NFL helmet liner, which evolved from first files to an on-field product in a matter of 8 months. In 2020, the NFL rated Riddell's SpeedFlex PrecisionFit Diamond as the top-rated American football helmet for player safety [1]. Watching Sunday football and cheering on the professional athletes who are wearing the helmet on national television is indescribably exciting and (at times) the highlight of some games for me (while I'm still in the process of becoming a big football fan, I have always been a big fan of innovation and safety)!

Since then, I've created and grown Carbon's Business Development organization. Riddell was the tip of the spear and allowed me to provide a compelling vision to company leadership: our customers needed *more* than world-class 3D printing technology and design tools – they needed a white-glove Sherpa to guide them through the inevitable difficulties of implementing a new design method, manufacturing technology, business model, and global supply chain. With my engineering and consulting background, I was confident I could hire, train, and grow an in-house team that would serve our customers with DLS™ expertise and product development best practices.

Today I lead a team of extraordinary business directors and design engineers who engage with highly innovative OEM research and design teams to transform their products and businesses using Carbon DLS™. In reaching out to highly capable product leaders, we often find customers who are thrilled at the idea of incorporating 3D printing at scale and achieving unprecedented levels of innovation. However, in opening Pandora's box of endless design opportunities, customers are faced with strategic and broad-stroke questions that can be paralyzing or stifling. Some customers are frozen by the uncertainty, while others spend months chasing research & development (R&D) ideas that never come to fruition – failing to get approval and buy-in from their leadership team to pursue business.

This is where my team steps in, and I absolutely love my team! Our shared passion for innovation and competitive drive for excellence enables us to confidently lead customers through ambiguity and change. Team members are pushed on a daily basis to balance the proverbial tightrope of art and science. They have equal amounts of guts and confidence to pursue their intuitions based on previous (sometimes unrelated) successes and failures. The engineers follow the systematic steps of the scientific method; with a logical, hypothesis-based approach, they ask tough questions, fearlessly dive into unknowns, find trends, prove theories, and relentlessly support customers in evolving obscure concepts into mature products.

So… What Is It that You Do Here, Exactly?

I'm often asked the question, *"is your team an internal design house?"* Well, not exactly. While we heavily focus on the design phases of a product's lifecycle, our objective and scope are much broader than helping customers design parts. We search for and target strategic industries and customers that can benefit from the added value of DLS™. Once we pick and prioritize a key customer, we lead them through the implementation of an innovative technology, unorthodox business model, and new supply chain to set unprecedented standards for their respective products. Beyond creating innovative designs, my team educates, consults, troubleshoots, analyzes, synthesizes, and prints, then once an OEM is ready for scaled production, my team facilitates relationships with and enables a smooth transition to global contract manufacturers using DLS™ for a holistic end-to-end solution.

We're pushing 3D printing to be more than just making the unmakeable – it starts with creating the un-designable. By unlocking an infinite combination of structures, textures, response profiles, and characteristics, our customers are provided with an array of solutions within one technology suite. The ability to fine-tune a variety of features using a single platform enables customers to create bespoke, or customer made, products at scale, while meeting unique product requirements and unit economics. A tailored resin-lattice combination can transform a product that formerly used separate pieces of foam for impact absorption, rotational energy dissipation, and comfort into an integrated energy management system. And, customers are able to differentiate their products across competitors because they have significantly more control over how each piece looks, feels, and performs.

At the intersection of hardware and software, it's imperative that our product development approach is balanced accordingly. With cloud-based algorithms that can create an infinite number of simulations and final-finish parts that can be used off the printer, we have the ability to work through quick sprints and use an Agile approach for testing and incorporating feedback. However, given that we're creating physical components or sub-assemblies, we cannot avoid critical path timelines driven by typical supply chain requirements, product testing, or certification. As much as we want to break free of traditional Waterfall methodologies, there are practical limitations to which we must adhere.

My team remains engaged through each iteration, sprint, and print as a concept evolves using the Carbon platform. This is why choosing the right customer is so critical – at times it can feel hectic, disorganized, and borderline impossible, but a strong partner remains steady throughout the journey.

What Makes a Great Partner?

In the early existence of our relatively new team, it was difficult to pinpoint exactly why development programs took off with select customers and not others. I'd often hear my team say, "It just works" or "Our teams get along really well," and while these were accurate statements, they lacked the details to help us crisply identify promising prospects.

After several "lessons learned" and post-mortem meetings, the team has gathered a list of qualities that have distinguished our most visionary and successful partnerships. In our experience, great partners are ones who:

– Embody bold and purposeful values and goals, which are aligned with those of Carbon.
– Have a compelling vision for transforming their products or business model.
– Understand their end-user so well that they anticipate their needs and wants – and can articulate them succinctly.
– Define and prioritize quantitative and qualitative success criteria early in the design process – and relentlessly stick to them through launch.
– Mobilize and expedite team engagement as they recognize there is an inherent competition with time to be the first and best.
– Hypothesize, test, iterate, and adjust quickly or without difficulty.
– Focus on data-driven designs and share data openly.
– Remain engaged and show resilience when the inevitable unknowns or problems arise.

It takes a few conversations or workshops before we recognize whether a customer has the grit to succeed and the passion for a product with compelling product-market fit. Sometimes we find answers organically, but more often than not, we make deliberate efforts to gain insights and create a profile of our partners as quickly as possible. This enables our team (and the rest of the company) to feel confident that the portfolio which we are pursuing will have a strong return on investment. With limited resources, time, and budget, it's imperative that we focus our attention on strategic partners and products with a high likelihood of success.

How do we get to know our customers? Here is a sampling of questions that we ask early in engagements with new partners:
How did you hear about Carbon?
How do you see/position yourselves in your market(s)?
What are you interested in improving? What currently works, what doesn't?
What problems are you trying to solve?
How would you like to differentiate your products?
Who are your biggest competitors?
What design capabilities do you have?
What is your current business model? (i.e., direct to customer, retail or through labs/medical?)
What do your manufacturing and supply chain operations look like today?

As we nurture and grow in a relationship with a new customer, our objectives go beyond understanding them – we need to understand *their* customer as well. We take time to learn and embrace each customer's value drivers to ensure we are creating products their end-users will love. Whether setting a new standard of performance or eye-catching differentiation, we focus on premium customers and end-users who value what we thoughtfully craft.

Create the Sandbox, Iterate and Evolve

Finding customers who know exactly what they want to do is rare. We all know the famous quote from Henry Ford, *"if I had asked people what they wanted, they would have said faster horses."* Some customers *think* they (or their end-users) know what they want, but after a few iterations, we discover critical feedback or information that was not communicated initially. Alternatively, customers may acknowledge they don't know where to start, so they have a laundry list of applications that they want to improve using DLS™.

Whether or not an eager customer has pre-selected a component they would like to commercialize, we are cautious to not hastily rush to production. Our team efficiently helps the customer down-select the best candidates and ultimately pick one debut product to revolutionize their respective category or industry. This helps the team stay focused, move quickly, and learn within a limited set of boundaries. Ideally, we rinse and repeat as soon as the first product is launched to leverage the best practices we have established and more efficiently launch additional products in even shorter time frames.

As the team progresses in the early conceptual phase, we focus on defining and evolving key product requirements. Instead of asking a customer exactly what they want, we repeatedly ask one key question: *"what's the sandbox?"* In other words, what are the physical, economical, manufacturing, supply chain, and consumer constraints that define what a product is (and equally important, *is not*). Throughout the development lifecycle, the sandbox where we "play" inevitably expands, contracts, and evolves. Our team has flexed and built the necessary muscles to feel comfortable working through a state of constant change and ambiguity.

Through a series of design workshops, our design team engages with OEM R&D or product teams to gain insight into an ideal concept. Here are some sample questions we ask throughout those meetings:
What type of energy are you looking to control? (i.e., *impact absorption or energy return) (Fig. 3.2).*

Fig. 3.2 Material-lattice configurations can be designed to absorb impact (left) or return energy (right)

What specifications does the product need to meet?
Are any certifications required for commercialization?
What aesthetic features does an ideal product include?
Will the product be designed to fit a custom shape?
Who is your target end-user?
How big is the target market?
How much market share do your current products have?
What are the volume-price trends for your products?
How do you define a minimum viable product?
When is the target launch date?

As a product evolves from initial concept to final design, we take a consultative approach to support customers in their decisions as they implement a new technology and process. We provide baseline options and compare variations (aka "trade-studies") that bookend their requests, then ask for quantitative and qualitative feedback to better understand what they do and don't want. Examples of tradeoffs may include print time, cost, throughput, surface finish, weight, or structural complexity. Providing customers with options and comparisons allows them to more clearly articulate their must-haves vs nice-to-haves, accelerate the product development lifecycle, remain engaged, and reduce churn.

I have often related the significance of trade studies to visiting an eye doctor. When getting glasses, the optometrist does not ask, *"how well do you want to see?"* Instead, they ask, *"which is better? Option 1 or Option 2?"* and repeat the process until diagnosing a prescription. This is a subjective and self-correcting process that empowers each individual (or customer) to crisply see and articulate details that previously could not be envisioned. As we print, iterate, and repeat, our team empowers customers with a streamlined process to hone in on an ideal solution quickly and effectively – one that is entirely based on their unique product vision.

We take great care in listening to our customers and fine-tuning products specifically to their needs. Because of this, two customers in the same industry could create significantly different products, as they have unique target consumers, goals, and desired features. Unlike other manufacturing methods, Carbon DLS™ enables OEMs to use the same technology platform while still creating unique and differentiated results.

Case Study: CCM Custom and Retail Helmets

CCM is a hockey equipment manufacturer based out of Montreal, Canada. For over 100 years, CCM has been leading the hockey market with industry-first innovations in hockey skates, helmets, and protective equipment [2]. CCM chose to partner with Carbon to explore how to leverage DLS™ to revolutionize hockey protective equipment for athletes at all levels. In 2020, our companies announced a partnership for the launch of two new helmets: Super Tacks X with NEST Tech [3], a custom-fit helmet for NHL athletes, and a retail helmet for players all over the world (Fig. 3.3).

My team began an engagement with the CCM research and product development teams in 2019 to assess a portfolio of potential next-generation products to incorporate Carbon's technology. The combined CCM x Carbon team honed in on key elements that athletes commonly asked for in a best-in-class hockey helmet, including breathability, comfort, and protection. Leveraging their scanning technology used for custom skates, CCM created a scanning solution for head shapes to drive a custom helmet design.

CCM's inputs, feedback, test results, and player insights drove each iteration to be efficient and meaningful. The Carbon team, including cross-functional experts from various engineering, software, materials, and process manufacturing teams,

Fig. 3.3 CCM SuperTacks
X with NEST Tech helmet

was able to accelerate design and test iterations faster than typical design cycles within the hockey industry. Throughout each iteration, the collective team prioritized player safety and how to precisely position and tune struts to absorb and disperse energy from various potential impacts in the rink.

NEST Tech stands for Next Evolution Sense Technology and several key attributes distinguish this product from any other hockey helmet on the market. The open lattice architecture allows air to flow throughout the helmet, keeping athletes' heads cool and hockey senses heightened. Printed with high accuracy and over 130,000 struts using the Carbon DLS™ process, the helmet is designed and printed to a custom scan, with a fit and feel unlike anything experienced before.

By the end of the 2019–2020 NHL season, 3 players were wearing custom helmets that were digitally designed and manufactured based on their head scan. Since the official launch of the helmet, production has been quickly ramping up to support demands for broad adoption during the 2021 season.

At the same time the custom helmet was launched, CCM also debuted a retail helmet, which will be available in 2021, offered in standard sizes to players around the world. The retail helmet design leveraged the knowledge gained from the custom helmet, with similar principles for superior design. The CCM × Carbon team was able to work quickly and diligently to complete the retail helmet design in less

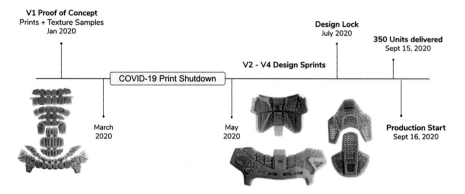

Fig. 3.4 Timeline of CCM retail helmet development

than 6 months – inclusive of COVID-19 delays! Yes, the team delivered through unprecedented hurdles and circumstances, creatively finding ways to engage in virtual meetings, execute design iterations, print, test, and provide feedback as soon as they were able to have limited operations within Carbon's and CCM's facilities. This was a true testament to the power of digital manufacturing.

The development timeline in Fig. 3.4 is representative of the accelerated pace of play when two partners are operating as one team, working toward a collective goal and efficiently using digital manufacturing to its full potential – against all odds in a global pandemic. The figure also includes the initial and final views of the NEST Tech components inside the CCM helmet.

Beyond helmets, we are motivated to continue bringing improvements to hockey equipment for years to come by enabling all athletes to get an edge and improve their performance while using CCM products with NEST tech.

Where Will We Go from Here?

"Consumers don't question why the same foam is used in two competitors' products-- why would they question if two products use the same 3D printing technology, which enables each customer infinite differentiation?"

My team is constantly evaluating new opportunities and investigating where we could revolutionize categories with the right partners. We're continuously pulling at little strings and watching new opportunities unravel within various segments of helmets, seating, gloves, apparel, and beyond.

Moving forward, I expect that more and more companies within the same industries will be incorporating DLS™ as we become the new standard for elite performance and protection. Consumers don't question why the same foam is used in two competitors' products – why would they question if two products use the same 3D printing technology, which enables each customer infinite differentiation?

Whether looking to level-up products using custom solutions, superior performance, improved comfort, or a mix thereof, our partners are embracing digital manufacturing with unprecedented results. The Carbon technology platform enables customers to design and make parts faster than any traditional form of manufacturing, and I am thrilled to continue supporting customers in accomplishing their unique product goals!

Closing Thoughts and Personal "Lessons Learned"

I couldn't write a chapter contributing to the accomplishments of women in 3D printing without providing perspective and reflection, with the hopes that I would encourage another rising leader in her field. In terms of professional development for the woman in additive manufacturing (or any traditionally male-dominated field), my personal advice to you:

Find Champions of Positivity and Stay Close to Them Traditionally the word "champion" is a workplace term that refers to an executive with a position of authority and influence, who advocates for your success. However, I've embraced an expansion of the term to include any subordinate, colleague, mentor, or executive dedicated to and encouraging positivity and success of an individual or team. This doesn't necessarily mean finding people that think similarly – some of my best supporters and mentors have very different ways of thinking. But in those differences, they consistently remain steadfast in appreciating my perspective and making me feel valued. I've found myself to be most confident and fulfilled by work when I'm surrounded by individuals who are in my corner, openly sharing feedback in positive ways and advocating for our collective success, regardless of their point of view or rank at the company.

Manage 360° and Build Strong Foundations of Trust Throughout my career and business school, I've taken a number of courses on honing leadership and management skills. Of course, managing direct reports or influencing without authority is important for getting a team to climb a mountain; but a less common or traditionally taught leadership trait is successfully managing up to your boss *and* to her/his peers (because your boss' opinion of you is undoubtedly influenced by the opinion of their peers!). It's healthy and often helpful to have complimentary views or respectfully question why decisions are made. However, when doing that, it's important to help leaders understand how your recommendation for change or growth aligns with their team's vision, mission, or goals to better enable your runway for success. Building positive relationships, embracing their mission, communicating frequently and effectively, delivering on commitments, and anticipating the needs of cross-functional leadership will establish trust throughout organizations and payoff in spades.

Advocate for Yourself This is an area where women commonly struggle (myself included – even writing this chapter was a challenge! I had several people add comments to initial drafts that I had focused more on my team's accomplishments than my own). Touting what you've done well, where your gut led to a breakthrough, or where your dedication and discipline showed unshakeable grit, can feel uncomfortable and borderline obnoxious. Sometimes it's even discouraged. I've always found it easier and significantly more enjoyable to advocate for my peers (male or female) than for myself. At some point, though, I realized advocating for women began with advocating for myself. Over the years this has been as subtle as tactfully speaking my opinion, openly asking for increased responsibility or confidently pushing for a promotion. If this is an area where you struggle, seek out women who are respected and successful in their roles. Even the most intelligent and confident women have had to navigate this space and are likely willing to share tips and encourage your growth.

If you'd like, feel free to reach out to me at erika.in.3dp@gmail.com. I'm happy to be a source of encouragement in your career, answer questions on the topics in this chapter, or discuss how we could potentially partner to create parts at scale. Together we can continue to empower women in 3D Printing and find ways to innovate, push boundaries, and revolutionize products!

References

1. NFL: Helmet Laboratory Testing Results. NFL Player Health & Safety. https://www.playsmart-playsafe.com/resource/helmet-laboratory-testing-performance-results/. Accessed 1 December 2020 (2020)
2. CCM: Our Roots. https://ccmhockey.com/en/about-ccm. Accessed 1 December 2020 (2020)
3. CCM: CCM x Carbon. https://ccmhockey.com/en/ccmxcarbon. Accessed 1 December 2020 (2020)

Erika Berg is the Managing Director of Business Development at Carbon, Inc. After joining the company in 2017, she built and grew a new organization called Application Development, which focuses on partnering with customers who have a vision for innovation and a desire to incorporate Carbon Digital Light Synthesis (Carbon DLS™) process into their products at scale. From 2018 to 2020, the Application Development team launched over 30 new SKUs of products with ground-breaking customers like Riddell, Specialized, fi'zi:k, and CCM.

In 2018, Berg managed the development and launch of Riddell's SpeedFlex PrecisionFit Diamond liner system, which is a set of custom scanned, designed, and 3D-printed pads inside of the helmet. In 2019, Carbon's work with Riddell was recognized by the Additive Manufacturing Users Group (AMUG) as the winner of the Advanced Concept Award in AMUG's Technical Competition. In 2020, the NFL rated Riddell's SpeedFlex PrecisionFit Diamond as the best performing helmet for player safety and was used by hundreds of NFL and NCAA athletes in its first year of launch.

With 13 years of experience in product development and consulting services, Berg is recognized for leading cross-functional teams through ambiguity with focus and resolve. She began her career designing commercial jet aircraft at The Boeing Company, supporting the 787 Dreamliner and New Airplane Study families. Berg also worked in Boeing's product marketing team and was a selected participant in Boeing's 2-year Leadership Development Program for rising leaders.

After Boeing and prior to joining Carbon, Berg was a manager in Deloitte's Strategy & Operations Consulting practice, advising Fortune 500 companies on product development and supply chain best practices. She holds an MBA from the University of North Carolina at Chapel Hill and a Bachelor of Science in Aerospace Engineering from California Polytechnic State University, San Luis Obispo.

Chapter 4
A Champion for Additive

Stacey M DelVecchio

Introduction

"We are planning to increase our engagement in the area of 3D printing and are looking for someone to lead this effort. Would you be interested in this position?" I was asked this question by Bonnie Fetch, a woman I admired and someone who would eventually be my mentor, my advocate, and my manager. Bonnie was a director at my emoloyer where I had worked for the past 25 years, and I had no idea how to respond. It was 2014; I had no background in 3D printing. None. I was just coming off a special assignment, having spent the previous year serving as the president of the Society of Women Engineers at a global level. While the presidency was an incredible experience, it was time to get back into a more traditional role at work. Despite having no experience in 3D printing, I did have 25 years of component product development experience in engineering and manufacturing coupled with 15 years of engineering leadership. I had no idea if 3D printing was an area I wanted to pursue. Spoiler alert; I accepted the job. But my experience leading up to this point, what I did in this role, and what I did after the role are the subject of this chapter.

Leading Up to Additive Manufacturing

With a B.S in chemical engineering from the University of Cincinnati, I embarked on my career at a Fortune 50 company in 1989. Chemical engineering was not the typical degree employed at my new company, but I was hired to work on non-metallic components, mostly in the area of process design. The company

S. M DelVecchio (✉)
StaceyD Consulting, LLC, Peoria, IL, USA
e-mail: staceyd.in.3dp@gmail.com

© Springer Nature Switzerland AG 2021   33
S. M. DelVecchio (ed.), *Women in 3D Printing*, Women in Engineering and
Science, https://doi.org/10.1007/978-3-030-70736-1_4

manufactured a lot of their own non-metallic components at that time and found chemical engineers to be a good match for this non-traditional role. I was pleased to find what sounded like, and proved to be, an exciting job, especially since I had been working at a different company where I was not very interested in the work I was doing. Plus, my husband-to-be had found a job at the same company a few months earlier. Taking this new job allowed us to live in the same city, something that was important to us and would remain so until we retired from the company 30 years later.

I spent my first 10 years at the company working on the process development work I was hired for. But it wasn't just one component or one process. The work encompassed process development for numerous rubber and plastic components used on our equipment. After several years on the process development side, I moved into manufacturing support for process fluids, water, and paint which allowed me to see another view of the company. I no longer had product-specific responsibilities, but was concerned about the support processes to make equipment the company produced. In this role I also had the design, build, and start-up responsibilities for a brand-new facility in Tianjin, China. It was a role that was not only technically challenging but personally challenging as it meant I needed to spend a significant amount of time in China. To complement this manufacturing support experience, I worked as an operation supervisor to get the first-hand experience of getting parts out the door. It was invaluable experience, especially in regard to my front-line leadership skills.

As I moved into engineering leadership, I had several roles that continued to shape my leadership skills. I was chosen to be a 6 Sigma black belt in the first wave of black belts in the company. It was an honor to be named a "wave 1" black belt. The role not only gave me tools to solve problems using facts and data as the foundation, it also showed me first-hand the importance of relying on the subject matter experts in my team. Future assignments as an engineering manager in non-metallic components complemented what I had learned as a black belt and strengthened my leadership skills. It was in these roles that I really started to appreciate the value of components. I eventually moved from components to a role in engines, focusing on fuel systems and a brand-new engine platform. I was a chemical engineer with no background in engines so I really had to rely on my leadership skills, my network, and especially my team's technical skills. I learned valuable life lessons from being in a leadership position without the technical expertise, something that I later would apply to my additive manufacturing assignment.

Throughout these various roles, I was active in the Society of Women Engineers (SWE). Based on my passion for the mission of SWE, "to empower women to achieve full potential in careers as engineers and leaders, expand the image of the engineering and technology professions as a positive force in improving the quality of life, and demonstrate the value of diversity and inclusion" [1], I chose to run for the highest office of Society president. With the backing of my company, I went on a special assignment during my term as the president, knowing that I would have a job to return to after my term, but unclear what the specific role might be.

While I had no grand plan for where all these jobs were taking me, I did enjoy the work and the company. Each new role added to my expertise in the value proposition of components at the company and my personal network in the engineering community. These areas would be key for my eventual role in additive manufacturing.

Getting Started

I accepted the job from my director to lead the additive manufacturing efforts at the company in 2014. I was charged with developing a strategy, and then implementing it. But where does one even start with such a grand objective with relatively little background and guidance? I had the backing of my vice president, my director, and my immediate manager, who all supported the division's mantra to "think big; start small; act fast." With so many years at the company, thinking big was not a problem. We were a large global company that made huge equipment. We owned big. And starting small wasn't too much of a stretch either. We were a 6 Sigma company that also had a rigorous new product introduction process, both of which saw the value in pilots and prototypes. But act fast? I knew that was going to be a challenge, especially in a company like ours that had been around since 1925.

There were a few general ideas that our leadership team had in mind that I embedded in our path forward.

- My division held design control for the company's components. In this capacity we would be the leaders in the early adoption of additive for the enterprise.
- We would build a lab, which we would call the Additive Manufacturing Factory, where we would print production parts, develop our printing processes, and conduct our research.
- We would be the enterprise technical experts for additive.

My first step was to get myself up to speed on the technology and start to build a team. Based on the counsel of the internal experts we had, I attended my first Additive Manufacturing User Group (AMUG) conference. The conference opened my eyes to capabilities and the opportunities available at that time. I quickly realized that the supply base wasn't ready for the vision we were creating within the company; the only way it would be possible was if we engaged the broader supply base. Despite numerous changes personally and professionally in the coming years, I would remain a strong advocate for the industry training opportunities like AMUG and the RAPID +TCT Conference, which is North America's most influential additive manufacturing conference, produced by SME. My attendance at these conferences made a significant impact on me as I realized the need for end users, like myself, to be engaged in the industry, sharing what was possible without divulging any trade secrets.

To build the Additive Manufacturing Factory, I needed help and found just the right person to lead the effort. This new team member previously ran the company's Rapid Prototyping Lab and was one of the few people within the company that had years of additive experience. With his help, we identified what printers and support equipment we needed. I tackled the issue of where everything would be located. There was a lot of debate on whether or not we should locate the equipment within an existing corporate facility, close to one of those facilities, or someplace completely external. There was pressure to locate it remote from current facilities so we wouldn't be hindered by any 'business as usual' from the company. Being remote would, in theory, allow us to operate more like a startup. In the end, I chose to build the factory within an existing facility that was the manufacturing research building. Of course, moving to a research building was about as far as we could be from a startup, but I firmly believed it was important to not only be visible, but to be in a place where we would have access to numerous other manufacturing processes that would be needed for post processing. It also meant that we could co-locate with the engineers that had been doing research on metal printing, and the long-established Rapid Protyping Lab. In future years, I would continually question my decision to locate the Additive Manufacturing Factory in an existing building as it saddled us with a lot of period cost and facility processes. Overall, I still think it was the best place to be.

I continued to be pushed by leadership to act faster. Making significant building modifications and buying capital equipment were typically done within the company via very established processes. I was dealing with a project that was only an idea in a leader's mind, with a dollar amount attached to make it happen. That was it. With some of my new team, we made immediate decisions on what equipment was needed. We made quick decisions on what the Additive Manufacturing Factory would look like and how much space we needed. There is no doubt that given more time, we would have made different decisions. However, I'm not sure they would have been any better. With the speed we were asked to move, it certainly taught me a lesson in moving forward versus getting stuck in decision-making mode. My team I moved into our new factory just 15 months after I had been approached about the job. Our new factory included not only our new 3D printers, but also the ones previously dedicated to research and the printers that were moved from the officially close Rapid Prototyping Lab. In terms of what had been done previously at the company, this was lightning fast [2].

We, as a company, got the "big" part of my leaders "think big; start small; act fast" mantra; however, I continued to be challenged with thinking big enough. At one point early on, my executive called to ask if we were working with a company called Made In Space. This company, together with NASA Marshall Space Flight Center, had recently remotely operated a 3D printer to build the first parts ever made off-Earth. They were focused on printing in space. More accurately, they are focused on developing state-of-the-art space manufacturing technology to support exploration, national security, and sustainable space settlement [3]. I focused in on the "printing in space" descriptor and actually found this inquiry from my executive to be rather humorous. Why would we be interested in printing in space? I politely told him that we were not working with this company and went about business as usual. Months later, in recounting this inquiry to other more innovative people than me, they thought that of course we should be working with Made in Space. If we, as a society, were going to establish a settlement on Mars, we would need to mine for materials and build structures. Certainly, my company would be interested in these future ventures if they ever became a reality. Good point. I wasn't thinking big enough.

Flexibility

Being in charge of deploying an emerging technology that was still developing was a challenge for me. I wasn't as flexible as I'd thought I was. I needed to learn how to act on ideas or concepts, and then quickly redirect our work if those concepts weren't successful. This mode of piecing together something quickly to check out a concept was new to me, but it was something people in the maker community had been doing for years. I am not a maker. That meant more challenges for me personally. My personal work history was not in design; it was in process development, strategy, and deployment. This meant I needed to be flexible in all of these areas.

Over the course of a few years, my group would put together several innovative ideas to rapidly deploy additive manufacturing. These ideas were mostly effective and well executed, but several of them quickly lost their appeal, so we moved on. Our more innovative initiatives are highlighted below.

Nomadic Printers

The Nomadic Printer concept was to put industrial printers in the hands of the facilities or design groups at no charge to them. The printers were not desktop printers. These were higher end material extrusion printers, more commonly known as fused deposition modeling (FDM) printers, and polymer material jetting printers. The printers were purchased by my group and loaned out for 6 months at no charge to different facilities in the company. This agreement kept the facility hardware asset base and associated costs from being negatively impacted by having a printer. The only thing the facilities had to do was buy the consumable material, identify someone to learn how run the printer, use the printer, and share what they were learning. The printers were called nomads since they didn't have a permanent home. Some facilities were wildly excited about the program and kept their nomad running around the clock. Other facilities had no idea what to do with it, other than to print name tags. We quickly learned that the material jetting printers were too sensitive to be on a rotation program like this, so we brought them back to the Additive Manufacturing Factory for their permanent home [4]. We also learned that beyond the facilities that were wildly excited about the printers, we struggled to find people that wanted a nomad. While the nomad printers were a step-up from a desktop printer, they still weren't quite good enough to be making production parts. This limitation proved to be an issue for some facilities. After a couple years, we decided

Fig. 4.1 Craig Woodin and Jim LaHood preparing the first nomadic printer for delivery to a facility

the nomadic printers had run their course and it was time to end the program. We internally transferred the printers to those facilities that were interested (Fig. 4.1).

Additive Manufacturing Summit

I knew it was critical to educate a high percentage of the thousands of engineers at the company in the basics of additive, versus having a small group that would be responsible for everything. In addition to my own group, there were just a few groups scattered around the company that were having some great successes with additive. I also knew we weren't going to get hundreds of engineers to conferences like AMUG or RAPID. However, since we worked for a big company with thousands of engineers, we had the luxury of hosting our own summit. There was no charge for people to attend the summit. They only had to pay for their own travel costs from their department budget. We chose speakers from around the company to highlight their successes in additive. We also brought in some suppliers to showcase their equipment. In the second year of the summit, we added a "design for additive manufacturing workshop" to the agenda. The summit was definitely a success, training hundreds of engineers, but like everything, there was a cost to it, mostly in time. With a small group that was constantly pulled in multiple direction, we chose to transition to a summit that was not annual. We eventually merged the additive summit into a similar structure that was for manufacturing in general, but would return to the additive manufacturing summit as needed [4, 5].

Design Competitions

In the world of innovation, competitions are very common. We thought it would be a great idea to have our own design competitions that would be open only to our employees. We embarked on a plan for a quarterly competition with a goal of getting people to think about how they could use additive in a safe environment such as a competition. The winner got bragging rights and my group would print their design at no charge. Each competition focused on a different area such as jigs and fixtures or design for additive. As was typical with most of our initiatives, this one also had a relatively short life. The first couple of competitions saw good participation and ideas that eventually were put into practice, but as time went on, participation slowed. We seemed to only be reaching the same small group of early adopters that were scattered around the company and we never really gained the encouragement of engineering leadership. The engineers were losing interest, or so it seemed. We ultimately stopped the program [4–6].

Costing Models

When the additive manufacturing efforts were first launched, my very supportive vice president felt strongly that cost, and the corporate bureaucracy to justify the cost, was a roadblock to innovation. When I say cost, I'm talking about any and all kinds of costs: product cost, consumable material cost, capital costs, building costs, utilities costs, and any other cost you can think of. My vice president was not talking about disregarding all cost or being fiscally irresponsible. He wanted to encourage people to take a risk, fail fast, learn from the experience, and move on. And if the engineers were overly concerned with cost, it was definitely a roadblock. So, I dutifully had my group start to print for groups within the company at no charge. Big mistake. As soon as people knew we were doing this, they had us printing stupid things, like flat plates with holes in them. Removing the barrier of cost meant that people saw our 3D printing services as a way just to get something for free instead of using the correct manufacturing process. They weren't using the free service to foster innovation. We quickly moved to a more reasonable model of charging people only for the consumable material we were using. The material only charge was much more reasonable and a much lower cost than the actual cost to produce the part. However, as business conditions and organizations changed at the company, we were challenged to recover our costs and needed to move to charging what the part actually cost us to produce. While I didn't want to stifle innovation, I thought charging what the part cost was a good thing. Our engineers needed to know the true cost of the additive manufacturing process to be sure they were using it in the right applications. We continued to refine how we did the costing, but thankfully never went back to the printing for free or for material cost only [5].

Additive Manufacturing Factory

With the speed we designed and built our Additive Manufacturing Factory, we needed to quickly select the 3D printers we purchased. I was pleased with what we had originally specified, with the exception of our metal printer. We purchased a specific model as another division within the company already had that model and we thought it would be good to standardize on it. We also had a smaller version of the machine already being used for research which reinforced it was a good decision. Shortly after we installed the machine, it was moved into the supplier's legacy line with little service support. We struggled to keep it running and to get quality parts off the printer. As a company, we weren't used to buying capital equipment that used technology that was still evolving, and evolving at a rapid rate. We made the brave decision to get rid of the printer and buy one that was more appropriate to our needs. The fact that we were able to do this encouraged me as it reinforced that our leadership was supporting some non-traditional ways of doing business that were needed when working with emerging technologies [2], [7].

Organizational Structure

The other major component that necessitated flexibility was the number of organizational changes that were happening at the company during this time. The Additive Manufacturing Team was moved from the component group to the innovation group, to the digital group, and finally to the research group, all in the span of three short years. And these weren't minor changes. They all involved completely different divisions with new executives and new leaders to bring up to speed on the program. Some of the groups didn't have any engineers except for my small team. Some of the groups were brand new and others had been around for decades. I spent hours educating my new leaders on the basics of additive, what we had done to date, and what our goals were. Each new set of leaders had a different vision on how the additive group would fit into their existing organization and how we had to adapt. The level of organizational change was a challenge and stretched me to the limits on the need to be flexible.

Engaging the Masses

Similar to my reasoning for hosting our own Additive Manufacturing Summit, I knew it was important to reach out to the masses, both inside and outside the company, in order to accomplish what we needed. Our outreach plan had three components: educate the engineers; educate the engineering leadership; engage outside partners. Getting the engineers educated and engaged was important, but if their leadership didn't understand what we were doing and why, it would be difficult for the engineers to move forward. Engaging organizations outside the company, be they suppliers, universities or not-for-profits was also a way for us to accomplish more than we would in isolation.

Engineers

There were a few divisions within the company that were doing some advanced work in additive. Our plan was not to compete with them but to complement their work. The trick was to engage the engineers without demotivating those that were on the right track. So, in addition to our summit, I identified leads for each product group. These leads were to be the first point of contact for each product group and they were to reach out to their engineering groups to assist with general education of additive and to assist in finding use cases. This structure had some success but wasn't as beneficial as I had hoped. As we worked with the groups to help identify use cases, engineering expertise in the specific product groups, be it components or vehicles, was lacking in my group. While my plan was for this to be a collaborative relationship, where my group had the expertise in additive and the product groups had the expertise in their product, it was at times difficult to pull the two together. I also felt strongly that it was important to use existing engineering systems for additive. Once we had approved a part to be used for additive, we needed to be able to tell other people in the company, from organizations like purchasing and manufacturing, that it was an approved 3D printed part. We found the best way to do this was by writing engineering specifications for the 3D printed material and including these specifications on the print.

Engineering and Supply Chain Leadership

At a higher level, we reached out to the engineering directors to educate them about the additive manufacturing value proposition within our specific industry. There had been, and still is, a lot of hype about additive quickly replacing traditional processes, especially in regard to the aerospace and medical industries. The heavy equipment and power industry, of which my company was a part of, is different than these industries which means our value proposition was not as obvious. We found we had multiple value propositions which were very dependent on the application

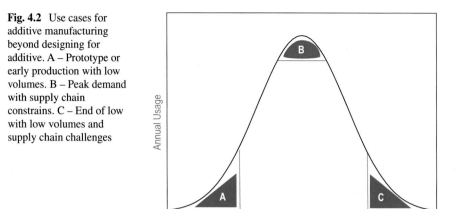

Fig. 4.2 Use cases for additive manufacturing beyond designing for additive. A – Prototype or early production with low volumes. B – Peak demand with supply chain constrains. C – End of low with low volumes and supply chain challenges

and where it was in its life cycle. Depending on whether the components were in research, new product introduction, production, or end-of-life made a big difference in the value additive could bring.

New applications could be designed to combine parts or to add intricate cooling passages for performance benefits, making the part financially worthwhile to print. It was fairly straightforward to see the value of printing early in the life cycle (area A in Fig. 4.2), especially if we were in the prototyping phase. We were already doing a little bit of this. But using additive as a way to complement our supply chain during peak demand (area B in Fig. 4.2) was a new concept. We struggled to get beyond the conceptual phase here as component validation was an issue. An ideal option to address the validation issue would be to validate a part that was produced with 3D printing and traditional methods at the beginning of the part life cycle. But at this point in time, the dual validation was cost prohibitive. We did understand the value of printing in the tail end of the life cycle (area C in Fig. 4.2) for our aftermarket business [8, 9]. With a strong brand and customer pull for parts needed late in the life cycle, we would often have supply chain challenges with the older parts. Additive manufacturing offered an option to supply the customer at times when no other option was available. Getting our leadership to see all these different channels for the effective use of additive, and getting our systems to facilitate this, were the challenges.

Outside Partners

The need to engage outside partners was clear. We would be able to accomplish more and enlist the knowledge of experts in the field. We not only worked on specific projects with national collaborative organizations such as America Makes, but we also hired consultants. With a well-established supply base, our challenge was to bring existing suppliers along in the journey versus solely adding new suppliers.

I'm not sure we ever mastered the art of how to do this. We also had several dealers who were very interested in additive and wanted to know how they could engage.

While the entities above were what I would call partners, there was also the overall industry and what could be accomplished through general public knowledge. If people knew what we were driving toward with additive manufacturing, I was convinced it would help the overall industry move forward. Having just finished my term as the president of SWE, I had learned a lot about the value of exposure in social media, press media, and external publications. When we first started on our additive journey, my leadership thought my experiences in this area would be a good skill set to leverage. Having people see our company as an early adopter of an emerging technology like additive manufacturing would be good for recruiting new talent and would reinforce our brand of being a technical company. I readily accepted interview opportunities by trade magazines and encouraged my group to share non-confidential information on what we were doing with additive. We also intentionally worked to be speakers at most of the major additive conferences, once again, reinforcing the work we were doing. As leadership changed, this willingness to be so open changed with it, but in those early days, it was a concerted effort to share our work.

Tours

One of the easiest, and most lasting items we did was to wholeheartedly accept tour requests. We did tours with engineering groups, new employees, budding engineers, all levels of leadership, dealers, customers…you name it…we invited them in. Each year we hosted thousands of people for tours of the Additive Manufacturing Factory. The tours would frequently be customized based on the audience, but that was fairly easy to do. My team knew what information was sharable to other groups inside the company as well as external to the company, so confidentiality was maintained. With so many people touring the facility, the group took more pride in their work place and kept it 'tour ready' at all times, well…almost 'tour ready'. It was a nice side benefit.

Leadership Support

I cannot say enough about the need for strong supportive leadership when launching an emerging technology. Of course, this statement could be used in just about any situation, but it's especially important when you're trying to get a large organization to change. Deploying an emerging technology is a challenging goal, but doing it in a company nearly a century old, like the one I was working for, was even more challenging. When I first took on the role, there were multiple levels above me that were supportive. They were interested in the work my group and I were doing. They wanted us to go faster. And they wanted to know how they could help. This support was in the inception when we were part of the component design group.

Less than a year after accepting my position and the creation of the group, we were moved to a completely new division centered around innovation and emerging technologies. While this may sound like a perfect fit, we lost our direct connection with the component design control. We still had the personal relationships from years of working with the component teams, but we were no longer part of the component engineering. On the positive side, we were once again with an incredibly supportive of group of leaders. My direct manager challenged us to push the boundaries of the corporate bureaucracy. He used to joke that he wanted us to get him in trouble. He was a brave leader that wasn't afraid to make unpopular decisions. I thrived in this environment.

In late 2015 the company encountered a downturn in business. We had weathered several of these in my 30 years at the company. But with this downturn, the company took major steps to restructure the company and right size some of the divisions [10]. My additive group was moved into the division that was responsible for digital products. We were the only group that had engineers in the entire division. There was an incredible amount of financial pressure on my new leaders, and here we were, a new group trying to implement an emerging technology. To say the two goals were mismatched would be an understatement. I firmly believed in what my small group was doing and did my best to adapt my group's goals. Eventually, the additive group moved to the very well-established research organization. The leadership in this group was stronger, although not as willing to take risks as what we had seen in the early days. While I was frustrated with the conservative nature of the group, it was helpful to have a leadership chain that was willing to establish attainable goals.

Within four short years, we had ridden a roller coaster of leadership support and styles. There is never a substitute for incredibly supportive, visionary leaders who are willing to push the boundaries. We had that leadership support in the beginning and would never have made the strides we had without their initial vision. It's a testament to company that we found our way back to strong leaders that could support the work, even though their conservative style was different than what we saw in those early days.

Moving On

After 5 years of leading the additive efforts at the company, it was time to move on to the next chapter in my life. There was still plenty to do in regard to additive and the work was still challenging, but my husband and I had been planning for several years to retire. We were ready financially and emotionally. I gave the company notice and they had my replacement identified in plenty of time for us to do an effective transition. I felt I was leaving the group in good hands, which is always a satisfying way to leave any role. I also knew I wasn't ready to completely retire from the field of additive, and quickly launched StaceyD Consulting, LLC. My intent was, and still is, to do consulting on the deployment of additive in the industrial space.

As I step back and reflect at this pivotal point in my life, I plan to embrace the following elements:

Stay Curious: I plan to continue learning, not only in the world of additive, but also in my personal growth. My time in additive manufacturing taught me the importance of pushing the boundaries, and when I do this, there's benefit of accomplishing more than I could otherwise. Those first leaders I had when I was in additive who wanted me to fail fast, learn from the experience and move on, had it right. I've carried that with me and will continue to do so. In order to fail fast, you have to move forward when things aren't perfect and that's ok. It's hard, but it's ok.

Inspire Others: I've had decades of inspiration from the people I've worked with at my company, my colleagues in the world of additive, my fellow members of SWE, and of course many family and friends. While I definitely continue to take inspiration from all of those around me, my goal is to pay it forward and inspire others, or at least aspire to inspire.

Do What Makes Me Happy: I realize I am incredibly privileged in order to make a statement like 'do what makes me happy', but this is where I am at this phase in life. Learning about advances in additive makes me happy; inspiring future leaders makes me happy; and getting plenty of sleep makes me happy too.

As I look to *stay curious, inspire others,* and *do what makes me* happy, I'd be delighted to be a resource for others. If you're interested in connecting with me regarding questions on this chapter, on 3D printing, or on the Society of Women Engineers, the door to my virtual office is always open at StaceyD.in.3DP@gmail.com.

Opinions expressed in this book are those of Stacey M DelVecchio, and do not reflect those of Caterpillar Inc., the Society of Women Engineers or SME.

References

1. Society of Women Engineers. www.swe.org. Access 16 Dec 2020
2. Caterpillar Announces Grand Opening of 3D Printing & Innovation Accelerator. https://www.caterpillar.com/en/news/caterpillarNews/innovation/caterpillar-announces-grand-opening-of-3d-printing.html. Accessed 16 Dec 2020
3. Made In Space. www.madeinspace.us. Accessed 16 Dec 2020
4. Molitch-Hou, M.: How Do You Integrate 3D Printing into a Big Business? Ask Caterpillar. https://www.engineering.com/3DPrinting/3DPrintingArticles/ArticleID/14678/How-Do-You-Integrate-3D-Printing-into-a-Big-Business-Ask-Caterpillar.aspx. Accessed 16 Dec 2020 (2017)
5. Langnau, L.: Caterpillar Inc.'s Plans for Additive Manufacturing. https://www.makepartsfast.com/caterpillar-inc-s-plans-additive-manufacturing. Accessed 16 Dec 2020 (2017)
6. Kuvin, B.: Caterpillar's grand AM plan. 3D. Metal Printing. **2**(2), 18–22 (2017) https://www.3dmpmag.com/magazine/article/Default.asp?/2017/5/3/Caterpillar%27s_Grand_AM_Plan
7. Griffiths, L.: Changing the world, one woman at a time. TCT Magazine North America. **3**(2), 28–29 (2017) https://www.tctmagazine.com/additive-manufacturing-3d-printing-news/changing-the-world-one-woman-additive-manufacturing/

8. Caterpillar Using 3D-Printing Parts on Heavy Equipment, With Little Fanfare. https://www.enr.com/articles/42654-caterpillar-using-3d-printed-parts-on-heavy-equipment-with-little-fanfare. Accessed 16 Dec 2020 (2017)
9. Crosby, E.: The nuts and bolts of 3D printing. CIM Magazine. **12**(6), 38–39 (2017)
10. Building For A Stronger Future, Caterpillar Announces Restructuring and Cost Reduction Plans. www.caterpillar.com/en/news/corporate-press-releases/h/building-for-a-stronger-future-caterpillar-announces-restructuring-and-cost-reduction-plans.html. Accessed 16 Dec 2020 (2015)

Stacey M DelVecchio is the president of StaceyD Consulting, focused on the business of additive manufacturing. She is a technical advisor with the Society of Manufacturing's Additive Manufacturing Community and has been an industry peer review panelist for Oakridge National Labs Manufacturing Development Facility. DelVecchio is a sought-after speaker on the value of additive and has been quoted in numerous technical magazines on 3D printing. Prior to launching her own consulting business, she was the additive manufacturing product manager for Caterpillar Inc. where her team leveraged the technology in all spaces, including new product introduction, supply chain, and operations. Her team focused on deploying the technology by working with Caterpillar product groups on design for additive. DelVecchio also managed Caterpillar's Additive Manufacturing Factory.

In her 30 years in industry, most of which was at Caterpillar, DelVecchio held numerous positions in engineering and manufacturing, including process and product development for nonmetallic components, production support for paint and process fluids, and the build and start-up for a green field facility in China. She is a certified 6 Sigma Black Belt, earning that classification by working on projects that included lean manufacturing, failure analysis, and employee engagement. Prior to her role in additive manufacturing, DelVecchio was the hose & coupling engineering manager with global responsibility for design, as well as the new product introduction manager for Cat Fuel Systems. She managed the project management office for Caterpillar engines with responsibility for the project management of all new product introduction programs, continuous improvement projects, and cost reduction project in the division. DelVecchio also developed an engineering pipeline strategy to ensure the best engineering talent was available to meet global enterprise needs.

DelVecchio holds a B.S. in chemical engineering from the University of Cincinnati in Ohio, USA. She was the 2013–2014 president of the Society of Women Engineers. She was named a fellow of the society in 2015 and received the society's Advocating Women in Engineering Award in 2018. DelVecchio was a vice-chair of the Women in Engineering Committee for the World Federation of Engineering Organization, representing the Engineering Societies of America and continues to champion global expansion of the Society of Women Engineers. She has spoken in a dozen countries on the value of gender diversity in engineering and is a strong advocate for women in engineering. In 2019, she published her first e-book, entitled *I Wish....* It is a quick read of a collection of her wishes for the engineering community based on her 30+ years as an engineering leader that happens to be a woman.

Chapter 5
Inkjet-Based 3D Printing: From Quantum Dots to Steel Tools

Amy Elliott

Inkjet-based additive manufacturing (AM) technologies are highly versatile among manufacturing processes due to their ability to shape any material in powder or ink form. The two inkjet-based AM technologies are binder jetting and material jetting, and as each of their names suggest, the former uses inkjet to deposit binder into the actual build material and the latter uses inkjet to deposit the actual build material. With each of these technologies comes unique opportunities in the form of embedding functional components within the build, creating material gradients within parts by selectively layering feedstocks, shaping non-weldable materials, and creating low-cost tooling with conformal cooling channels. Dr. Amy Elliott at Oak Ridge National Laboratory has explored a variety of materials and approaches with inkjet-based manufacturing technologies, focusing on industry-relevant applications such as injection molding tooling and mining drills.

Steels with Gradient Properties

At Oak Ridge National Laboratory (ORNL), Dr. Amy Elliott works to expand the usability of binder jetting technology through exploring new material systems and uses for the technology. Binder jetting works by spreading individual layers of powdered material and selectively binding them using an inkjet printhead. The printed parts can then be post-processed by either sintering them to full density or infiltrated with a secondary material, which is typically metal to create a metal matrix composite (MMC). Binder jetting is a technology that is especially attractive to the manufacturing industry due to its low cost of production enabled by its high throughput [2] (Fig. 5.1).

A. Elliott (✉)
Oak Ridge National Laboratory, Knoxville, TN, USA
e-mail: elliottam@ornl.gov

© Springer Nature Switzerland AG 2021
S. M. DelVecchio (ed.), *Women in 3D Printing*, Women in Engineering and
Science, https://doi.org/10.1007/978-3-030-70736-1_5

Fig. 5.1 Artifacts with gradient stoichiometries of TiC printed with binder jetting [1]

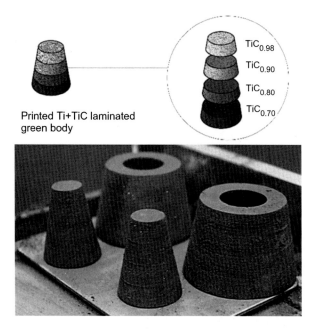

Printed Ti+TiC laminated green body

$TiC_{0.98}$

$TiC_{0.90}$

$TiC_{0.80}$

$TiC_{0.70}$

Multi-material printing is of interest to applications in extreme environments since equipment and tools used in these applications have competing demands on the material properties such as impact toughness vs. hardness. By printing with multiple materials, artifacts can be created that have properties that are tailored to the location within the part and ultimately tailored to the stress or environmental factors that the part experiences throughout its topology.

One way to accomplish multi-material printing with standard binder jetting equipment is by manually changing the build powder during the print, which allows for "2D freedom" in material selection. To accomplish a powder change during the print, the print is paused, the original powder is replaced with new powder in the powder supply hopper, and the print is resumed. Once the powders are placed, the printhead can bind these different powders together into a single part. From there, the post-processing of the part can be performed as normal: either with sintering to full density or infiltration with a lower-melting temperature material. Researchers at ORNL and Ben-Gurion University of the Negev teamed up to explore multi-material deposition by printing titanium carbide (TiC) artifacts with gradient carbon content [1]. TiC is commonly used in mining due to its high hardness and low density compared with materials like tungsten carbide. Further, when TiC is infiltrated with steel, some carbon migrates from the TiC to the steel matrix, which increases the steel's hardness. Thus, the purpose of this study was to explore the effect of the TiC stoichiometry on the infiltrated steel matrix while also demonstrating a gradient effect through multi-material printing with binder jetting. Four different powdered feedstocks of different titanium carbide (TiC) chemistries were utilized in the same print: $TiC_{0.7}$, $TiC_{0.8}$, $TiC_{0.9}$, and $TiC_{0.98}$ (Fig. 5.10). The printing was conducted on an

ExOne X1-Lab system, which has a build volume of 40 mm x 60 mm x 35 mm. To create a material gradient, each material was placed into the feed hopper of the binder jet printer to supply specific layers of the print. To accomplish this, the print was paused, the powder was removed from the feed supply via scooping and vacuuming, new powder was added to the feed reservoir, the power was leveled with the machine's roller mechanism, and the prints were resumed. By replacing the powder in the feed bin, the next layer that is spread is entirely of the new powder feedstock. This transition occurred 3 times throughout the print between the 4 different types of powder. When completed, the result was that the print volume was layered with 4 discrete sections of the different TiC chemistries and within the build volume the binder created bound artifacts within the layered construction. Unfortunately, due to the nature of removing parts from the powder bed, keeping the different powders from mixing during part excavation was impracticable, so unlike typical binder jet prints, the powder surrounding parts during printing cannot be reused. Thus, more work is needed in systems development to improve powder recyclability for multi-material powder beds.

Once the parts were printed, removed from the build volume, and the loose powder cleaned from each surface, the gradient carbides were infiltrated with steel to create a metal matrix composite. The infiltration was conducted under 10–4 torr vacuum with 1070 steel at 1400 °C for 4 hours and then at 1550 °C for 15 minutes. Once the samples were cooled, heat treatments were conducted at 900 °C for 30 minutes followed by oil quenching. This was performed in order to transform the pearlite microstructure in the steel matrix, which was next to the stoichiometric TiC ($TiC_{0.98}$), to martensite. The hardness was measured in each region of the four TiC powders, and was found that the change in hardness throughout the material corresponded directly with the gradient of carbon in the TiC blend sections [3]. This technology has application in mining and milling, where wear-resistance is a needed material trait. The gradient hardness will theoretically allow for more ductility toward the base of the "tooth" where the stresses are higher, while the tip of the tooth has higher hardness to accommodate the higher wear rates.

Binder Jetting of Highly Reactive Lanthanides

Advantages of binder jetting include the technology's flexibility with respect to powder composition and particle shape combined with the potential for large-scale production at low cost. Because of these advantages, Dr. Elliott and her team at ORNL have explored a wide variety of powdered material types with binder jetting along with industry partner collaboration. One of the most notable materials explored was a magnetic caloric (MC) powder that was printed in partnership with General Electric [4]. The salient property of the MC material (MCM) is that it can absorb thermal energy in the presence of a magnetic field, making MC materials highly valued in refrigeration research. Challenges arose in printing the MC material due to both particle composition and shape. The powder that was initially tested was highly irregular due to it being created via milling, which meant it tended to agglomerate during printing and affect the uniformity of the powder bed. More significantly, however, the penetration of the binder through the layers of MCM powder was radically different than any feedstock previously attempted. To establish process settings for a new material on a binder jet system, a few key parameters, including binder saturation, drying time between layers, and roller speed, were modified. To begin process setting development, the operator selected parameters that were most likely to succeed based on their experience. During the trial prints, the operator observed the print for over drying (witnessed by cracking and/or curling of the layers), under drying (witnessed by part shifting or smearing by the spreading roller), short spreading (shown as gaps at the end of the print bed where powder should be filling), and other characteristics. For some binder jet systems, the operators adjusted parameters on the fly until these issues were resolved and the print stabilized. Once the printing conditions were stable and the test print completes, the test print was cured, and the parts were depowdered or removed from the build of powder they were printed in. At this point, the operator was looking for "crisp" or sharp, well-defined features which would indicate that the binder saturation was at the proper level. Binder saturation is the amount of binder that the printer deposits into a given volume of the print, and when the saturation is too high, the part features are poorly defined and appear to be "swollen" and rounded. On the other hand, if the binder saturation is too low, the features may be well defined, but the part strength will be too low to be able to depowder them or handle them for post-processing. Part strength can be measured by printing ASTM rectangular bend bars measuring their strength using a 4 point bend testing setup [5]. When the saturation was adjusted, the print drying time needed adjusting as well, so several printing trials were needed to get each print parameter at a suitable setting.

For this work, samples were printed from the MCM granules supplied by Luvak Laboratories that were milled with steel media in an argon-filled container (Fig. 5.2).

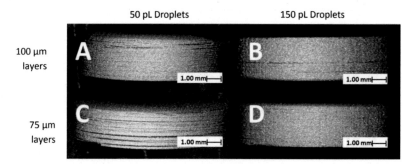

Side view of geometries printed from milled MCM powder with identical binder content but varying layer thicknesses and drop size. Coins in Figures A & B have 100 µm layers; Figures C & D have 75 µm layers. Coins in Figures A & C were created with 50 pL droplets while coins in Figures B & D were created with 150 pL droplets.

Fig. 5.2 Side view of geometries printed from milled MCM powder with identical binder content but varying layer thicknesses and drop size. Samples denoted by A & B have 100 um layers; samples C & D have 75 um layers. Samples A & C were created with 50pL droplets while samples in Figs. B & D were created with 150 pL droplets

The milled powder was sieved with 100, 200, and 325 meshes to remove larger particles. Initial printing trials revealed that the layers of each print would delaminate from one another, even when using the highest saturation settings. This phenomenon was attributed to the size and shape of the particles, which were highly irregular (i.e. jagged, non-spherical). We hypothesized that the irregular shapes of the particles caused a more tortuous path for the binder to penetrate when compared to the spherical powder that is typically printed with metal powder AM. The solution to this problem was to utilize a different printhead that provided larger drop sizes (150 picoliters vs. 50 picoliters) as well as decrease the layer thickness to the lowest setting (75 microns). The hypothesis as to why the larger droplets improved layer penetration was that the larger volume of fluid overcame the surface tension forces that were dominant for smaller droplets.

Although this work is very preliminary, it is a step toward additive manufacturing of MCM materials for refrigerating components. It is believed that developing this technology could increase the efficiency of refrigeration systems by removing the need for a compressor and other components [6]. Further, using binder jetting to shape the MCM materials will allow for complex heat exchange geometries that will further increase the technology's efficiency.

Neutron Collimators

Neutron imaging is a non-destructive characterization tool for revealing atomic structures and behaviors of materials in a wide variety of environments and conditions. Facilities like the Spallation Neutron Source (SNS) at Oak Ridge National Laboratory enable research and discoveries for researchers around the world but represent significant financial investment. As such, the ability to precisely define the neutron beam is of high importance; however, the hardware traditionally fabricated for neutron collimation is limited to simple shapes that can be cut and glued from pressed and sintered sheets of boron carbide- or other neutron-absorbing materials (Fig. 5.3). To overcome this limitation, researchers at ORNL's SNS and Manufacturing Demonstration Facility joined forces to investigate how a complex and custom-shaped collimator might be fabricated using additive manufacturing technologies. Dr. Amy Elliott was the manufacturing lead for the project, using her binder jetting expertise to aide in the production of collimator geometries. Toxicity of powdered material is a significant barrier to binder jet printing since current equipment has an open architecture that is not capable of fully isolating powders from the outside environment. Since the majority neutron-absorbing materials are toxic (e.g. mercury, gadolinium, boron, cadmium), non-toxic carbide and oxide-forms of these materials were explored, and boron carbide (BC) was selected as the "friendliest" material candidate. The initial attempts to print boron carbide were centered around BC-loaded thermoplastic filaments that were printed via extrusion AM. The issue with this approach was that the actual solids loading of BC was around 30–40%, but what was more significant was the amount of thermoplastic in

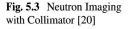

Fig. 5.3 Neutron Imaging
with Collimator [20]

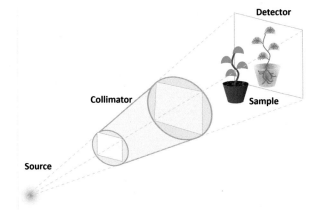

the composite. Thermoplastics are made of carbon, hydrogen, and oxygen polymers and monomers, and the hydrogen in these materials scatters the low-energy neutrons, impairing neutron image fidelity (Fig. 5.4).

Because of these drawbacks of extrusion 3D printing with particle-loaded thermoplastics, a solution was needed for shaping boron carbide with minimal polymer; binder jetting was selected for exploration [8]. In the binder jetting process, the binder solution that is jetted diffuses through the powder and remains at the necks or contact points between the particles, filling only a small fraction of the void space between the particles. As a result, binder jetting uses very little polymer to bind powders, making it the more ideal choice for shaping neutron collimating materials.

The research at ORNL in printing boron-carbide collimators centered around powder feedstock evaluation, print parameter development, and post-processing techniques. Using commercially available powder feedstocks is a challenge for AM technologies since AM equipment operates best with specialized powders. Ceramic powders like boron carbide are an even bigger challenge since they typically have poor flowability and packing. However, a boron-carbide powder was selected and print parameters were successfully identified. Post-processing techniques including infiltration with small amounts of polymer and melt-infiltration with aluminum to give the collimator better mechanical properties [9]. Preliminary experimentation revealed favorable neutron scattering properties, and the technology was licensed for commercial use by the ExOne Company [10].

Binder Jet Printing of Conformally Cooled Injection Molding Tooling

There's an adage in the field of AM that goes, "The top applications for additive manufacturing are tooling, tooling, tooling." There's a lot of truth to this mindset since manufacturing of tooling is typically expensive, and requires long lead times,

Fig. 5.4 Complex collimator geometry printed from boron carbide powder using binder jet [7]

Injection molding tool and finished cup   (Left to right: Molding core; sintered part, machined & polished)

Fig. 5.5 Printed and polished injection molding core installed in molding setup, including cup that was produced (left), and printed molding core cross section and polished work piece (right) [12]

and tools themselves are needed in low volumes and their geometries can be highly complex. Additive manufacturing can solve each of these issues since AM is best suited for low-volumes, complex geometries, and one does not have to wait on the procurement of a custom-size billet for each job since the printer uses the same feedstock no matter what the geometry [11].

Dr. Amy Elliott's team at Oak Ridge National Lab conducted a demonstration in partnership with an industry to show that medium-sized injection molding tooling could be produced with binder jetting using a single alloy (Fig. 5.5). The team utilized H13, which is a hot-work tooling alloy. The challenges associated with this endeavor included producing a preform with minimal defects and sintering the preform to full density with minimal distortion and cracking. Small defects in the print (which were a result of lack of maintenance on the machine) severely affected the success of the sintering. These defects were primarily cracks formed during sintering because of build errors such as short spreading or misplaced binder. Sintering of binder jet parts to full density is challenging because the preform is usually no more than 60% dense by volume, so the part has to shrink by up to 20% in each direction. This amount of shrinkage is rarely met without distortion or stress cracking since the material in the part will usually have thermal gradients. These gradients mean the outside of the part will reach the sintering temperature and densify before the inside of the part is able to reach the sintering temperature. As such, it is difficult to print larger parts and avoid cracking due to their size. Further, larger parts are more constrained on the crucible floor due to their weight and resulting frictional drag, so cracking and/or flaring of the geometry at the bottom is common. Finally, the weight of the part can induce a phenomenon known as "elephant footing," which is the swelling of the part at the bottom similar to when an elephant puts weight on its foot. To alleviate these issues, ceramic beads or rollers can be used to reduce drag of the part on the crucible, and elephant footing can be somewhat avoided if lower sintering temperatures are used; however, this is not always practical.

The experiment required numerous sintering trials to find a sintering cycle that would not cause significant distortion or cracking. After numerous attempts at printing and sintering, the team produced a part that reached 97% density with minimal cracking. The part was machined and polished to reach the needed tooling finish, and the tool was mounted on an injection molding setup by industry partner Innovate Manufacturing. They produced 100 cups with the printed tool. This demonstration constitutes the first attempt to produce a single-alloy part through printing with binder jetting and sintering to full density (i.e., no infiltration).

3D Printing a Tungsten-Carbide/Cobalt Rock Drill

Binder jetting is most well known for creating bronze-steel metal matrix composites (MMC's) via printing of steel powder and melt-infiltrating bronze to create a fully dense part. The method is highly advantageous since the melt-infiltration fills the porosity in the print rather than requiring shrinkage, which enables fully dense metal parts via binder jetting with full shape retention. Since the melt infiltration is critical to binder jetting for making fully dense metal-metal or metal-ceramic artifacts, Dr. Elliott and her team at ORNL have studied many different MMCs that can be created with melt infiltration. Typically, the two materials in an MMC pair will have different melting temperatures, will have similar coefficients of thermal expansion, and will have favorable wetting with each other once the matrix material has melted, among other properties.

In addition to being easily shaped with binder jetting followed by melt infiltration, MMCs have unique properties that are advantageous to many applications. A literature review of materials that can be created with pressureless melt infiltration revealed many material combinations with high utility, including tungsten carbide infiltrated with cobalt (WC-Co) [13, 14]. WC-Co is an especially useful material due to its high hardness and abrasion resistance from the WC combined with the high strength of the Co matrix. In mining and drilling, WC-Co is used as the main material for the tips of earth drills and even the body of the drill itself. With this value proposition in mind and the ability to create a WC-Co via melt infiltration of Co into WC, Dr. Elliott's team worked to manufacture a full-size rock drill by printing WC powder and melt-infiltrating Co.

Print process settings for the WC powder were developed for the ExOne MFlex system. A digital rock drill geometry was obtained and modified for the printing and infiltration experiment. The major modification that was made to the rock drill geometry was the addition of vertical bars or pillars underneath the blades that would theoretically allow for more even infiltration of Co during the furnace cycle and also support the blades during handling. The parameters of the project were to explore an infiltrated geometry much bigger than the rock drill, so a large trough was attached to the part print that would contain the cobalt metal.

Due to limited powder supply, a single print with two copies of the drill/trough combination was printed on the Mflex machine in one print. Since WC powder is highly abrasive, a hole was worn in one of junctions in the tubing for the powder delivery system (the system that retrieves overflow powder and moves it to a storage tank outside the machine). Other challenges arose during this printing trial, such as parameters needing to be adjusted throughout the build, most likely due to the fluctuating humidity in the facility where the printer is located. Once the print was complete, the build was moved to a curing oven, cured overnight, and allowed to cool for 24 hours. The parts were depowdered over the course of several hours, and the resulting artifacts can be seen in Fig. 5.6. Extra care was taken to excavate the parts due to the breathing hazards associated with WC powder. The resulting geometries had only a few defects most likely caused by humidity fluctuations. Namely, toward the middle of the build, some swelling occurred in the part and the columns that supported the rock bit blades. Regardless these build errors were considered minute enough to continue the experiment.

Once depowdered, the part was placed in a large, graphite crucible and the trough was filled with the amount of cobalt that corresponded to 70% of the WC print volume. Preliminary trials with around half this amount did not fully infiltration. The

Fig. 5.6 WC-Co infiltration (left) green part and (right) the part after sintering right before processing Co infiltration

Fig. 5.7 Final infiltration
experiment with distortion

bit was covered in alumina grit as an insulator and support during infiltration. The
parts were processed in 10^{-5} Torr vacuum to a temperature of 1600 °C for 6 hours
[15]. This temperature corresponds to the melting point of cobalt, and the hold time
corresponds to an estimated time for the temperature within the part to reach equi-
librium (based on previous work on this furnace) and to allow the molten cobalt to
fully infiltrate. Once cooled, the part was removed from the crucible and alumina
grit and inspected. The result of the furnace cycle showed some distortion of the
rock drill geometry that was not anticipated, as shown in Fig. 5.7. Until this time,
only bronze and steel had been attempted as an MCM pair, and this material system
had been proven scalable to very large parts on the order of 10 inches in each direction without distortion. The unexpected distortion that was encountered during this
large-scale infiltration was investigated, and it was found that the WC-Co material
system has a eutectic at the same WC:Co ratio that was being used in this experiment. Although project resources did not allow for further large-scale infiltrations,
subsequent studies showed that alternative processing cycles (e.g., pre-sintering the
WC to increase density and lowering the amount of Co for infiltration) solved the
distortion problem for smaller geometries [16].

Physical Unclonable Functions for Material Jet Parts

The ability to produce virtually any geometry using additive manufacturing has risen concerns in the area of counterfeiting of products. A way to combat counterfeiting is to place a security feature in the 3D printed part that would identify the part as valid. The security feature itself would need to be non-reproduceable by additive manufacturing, meaning the materials or geometry used would need to take advantage of randomness. Such a security feature is known as a Physical Unclonable Function (PUF) and can take many forms. Dr. Elliott worked at the Virginia Tech DREAMS (Design, Research, and Education for Additive Manufacturing) lab to create PUFs in polyjet prints by dispersing quantum dots (QD) in clear photopolymer that gets deposited with an inkjet printhead as seen in Fig. 5.8 [17]. The random dispersion of the particles creates a non-reproduceable fingerprint that could be registered in a database and referenced if the legitimacy of the part was in question. The quantum dots are a strategic particle since they are capable of shifting wavelengths of light (e.g., shifting UV light to the visible spectrum, aka fluorescing). This is critical for a security feature to have the challenge (UV light) and response (visible light) signals to be distinguishable from each other.

Quantum dots were functionalized (i.e., formulated with polymer ligands on the surfaces) to be dispensable in Objet VeroClear resin, and optical and rheological properties of the suspensions were characterized using a variety of techniques, including UV-Vis, UV rheometer, and others. Dr. Olga Ivanova (another researcher in 3D printing, currently employed at Open Additive) performed the functionalization and characterization of the quantum dots. The solution was deposited onto a film using a single nozzle inkjet printhead produced by MicroFab technologies in

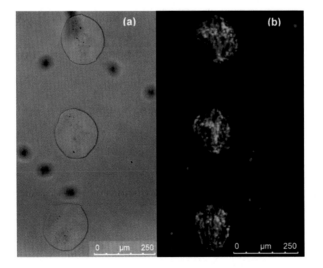

Fig. 5.8 Inkjet droplets of quantum dot loaded photopolymer deposited on a film in (**a**) visible light and (**b**) fluorescent light

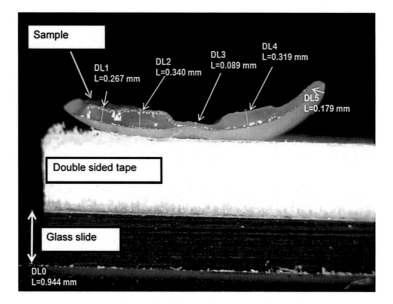

Fig. 5.9 Films for testing cure depth of quantum dot-loaded photopolymers

Plano Texas. The jetting of the suspensions was analyzed using the single-nozzle setup to ensure the addition of the quantum dots did not adversely affect the materials' ability to be deposited via inkjet. Individual drops were deposited, cured, and imaged using a fluorescent microscope [18].

The bulk of this research dealt with the effect of the quantum dots on the curing behavior of the photopolymer resin. Since the quantum dots themselves inherently absorbed UV light in the same wavelength that the photopolymer required for solidification, it was important to quantify the curing phenomena of the quantum-dot loaded resin. To quantify the effect on curing, suspensions of varied QD concentration were cured using varied exposure times, and the resulting film thicknesses were measured (Fig. 5.9). It was found that the quantum dots did impede the curing of the photopolymers to an extent that was proportionated to the concentration.

Material Jetting: Embedding Actuators

An advantage of building objects layer-by-layer with AM is access to every "voxel" of the part as it is being built. This access can be used to monitor temperatures, map defects, or even allow for the embedding of objects. Embedding is a strategy used in a lot of different types of 3D printing where a different material or even a functional object like a motor or actuator is placed within the print during the printing process (unlike traditional manufacturing where pieces are assembled after they are

Fig. 5.10 (**a**) CAD representation of finger, (**b**) built finger, (**c**) actuated finger [19]

shaped individually). The result is composite artifact with an object or material encapsulated in the main structure to produce a mechanical system. This approach has been used to produce a variety of mechanisms and composite structures, including actuated joints.

Polyjet is an interesting technology to explore with embedding since the polyjet process is already capable of depositing multiple materials into a single layer using the dot-matrix control of inkjet. Polyjet is a material jetting AM technology that utilizes an inkjet printhead to deposit photopolymers, which start in liquid form, get deposited with the inkjet printhead into the desired shape of the part layer, and are then solidified with a UV light. Using photopolymers of different durometers and the selective deposition of inkjet, a rigid structure with flexible joints can be fabricated. Combining this and the opportunity to embed fibers during the print, an actuated system can be fabricated. As such, Dr. Elliott of ORNL participated in a study using polyjet technology to print a specimen with a flexible joint in the middle. The embedded a polymer thread could be tensioned by moving a slider mechanism at the end of the device (Fig. 5.10).

The embedding procedure is similar across the different AM platforms. First, the designer must include an empty volume in print design that is the same geometry of the object to be embedded. For embedded objects that are not flat on the top (and cannot support the next layer of print), shape adapters can be printed beforehand and embedded with the object to create the needed support. During the print, the operator allows the print to proceed until the top of the empty volume is reached, in other words, before the print closes around the volume that was left open in the design file. At that point, the print is paused, any support material that was deposited into the void is removed, the object is embedded, and the print is resumed. Some complications that can arise during the embedding procedure are accidently marring of the print plane, dislodging of the printed part itself, or poor embedding strategies that lead the embedded object to interfering with the layers being printed above it. For wire embedding, a narrow channel is left open during the part design. The shape-memory alloy (SMA) wire that was used in later version of this work required special fixturing and channel design to keep the wire from bending above the print plane. Overall, embedding of actuators and other artifacts with 3D printed parts is a popular method for adding functionality to AM parts.

Conclusion

Inkjet-based additive manufacturing is a new area of research with many opportunities in terms of materials research and mechanical design. As a woman in 3D printing manufacturing research, Dr. Amy Elliott of Oak Ridge National Lab has forged a path in inkjet-based AM research centered around exploring new material systems, feedstock shapes, and embedded artifacts for adding functionality.

References

1. Levy, A., Miriyev, A., Elliott, A., Babu, S.S., Frage, N.: Additive manufacturing of complex-shaped graded TiC/steel composites. Mater. Des. **118**, 198–203 (2017)
2. Elliott, A. M., Love, L.J.: Operator burden in metal additive manufacturing. In *27th Annual International Solid Freeform Fabrication Symposium*, 2016, pp. 1890–1899
3. Levy, A., Miriyev, A., Elliott, A., Babu, S.: Additive manufacturing of complex-shape graded TiC-steel composites. In: 41st International Conference and Expo on Advanced Ceramics and Composites (2017)
4. Elliott, A., Benedict, M., Momen, A.: Additive manufacturing of highly reactive lanthanides. In *TMS Annual Meeting & Exhibition*, 2018
5. Baratta, F.I.: Requirements for flexure testing of brittle materials. In *ASTM Special Technical Publication*, 1984
6. Schroeder, M.G., Benedict, M.A., Momen, A.M., Elliott, A.M., Kiggans J.O., Jr.: "Method for Forming a Caloric Regenerator." 2018. https://www.ornl.gov/technology/201703848
7. Shoemaker, Sara. 2020. "ExOne Licenses ORNL Method to 3D Print Components for Refined Neutron Scattering." *ORNL News*. June 22, 2020. https://www.ornl.gov/news/exone-licenses-ornl-method-3d-print-components-refined-neutron-scattering
8. Stone, M.B., Siddel, D.H., Elliott, A.M., Anderson, D., Abernathy, D.L.: Characterization of plastic and boron carbide additive manufactured neutron collimators. Rev. Sci. Instrum. **88**(12) (2017)
9. Halverson, D.C., Pyzik, A.J., Aksay, I.A., Snowden, W.E.: Processing of Boron Carbide-Aluminum Composites. *Journal of the American Ceramic Society* **72**(5):775–780 (1989). https://doi.org/10.1111/j.1151-2916.1989.tb06216.x
10. Cramer, C.L., Elliott, A.M., Kiggans, J.O., Haberl, B., Anderson, D.C.: Processing of complex-shaped collimators made via binder jet additive manufacturing of B4C and pressureless melt infiltration of Al. Mater. Des. **180**, 1–9 (2019) https://vtechworks.lib.vt.edu/bitstream/handle/10919/46632/Elliott_AM_D_2014.pdf?sequence=1&isAllowed=y
11. A. Leandri: What can additive manufacturing do for tooling?, 3DPrint.com, 2015. [Online]. Available:. https://3dprint.com/55676/additive-manufacturing-tooling/
12. "Oak Ridge National Laboratory and ExOne Case Study." ExOne.com
13. Elliott, A., Nandwana, P., Shackleford, C., Waters, C.: Roadmap for metal hybrids net-shaped via binder jet additive manufacturing. In: MS&T (2017)
14. Shackleford, C., Arnold, J., Nandwana, P., Elliott, A.M., Waters, C.K.: Metal matrix composites formed by titanium carbide and aluminum net shaped via binder jetting. In: In *28th Annual International Solid Freeform Fabrication Symposium* (2017)
15. Cramer, C.L., Nandwana, P., Lowden, R.A., Elliott, A.M.: Infiltration studies of additive manufacture of WC with Co using binder jetting and pressureless melt method. Addit. Manuf. **28**, 333–343 (2019)
16. Cramer, C.L., Wieber, N.R., Aguirre, T.G., Lowden, R.A., Elliott, A.M.: Shape retention and infiltration height in complex WC-co parts made via binder jet of WC with subsequent Co melt infiltration. Addit. Manuf. **29**, 1–6 (2019)

17. Ivanova, O.S., Elliott, A.M., Campbell, T.A., Williams, C.B.: Unclonable security features for additive manufacturing. Addit. Manuf. **1–4**, 24–31 (2014)
18. Elliott, A.M.: The Effects of Quantum Dot Nanoparticles on the Polyjet Direct 3D Printing Process. Virginia Polytechnic Institute and State University (2014)
19. Stiltner, J.L., Elliott, A.M., Williams, C.B.: A method for creating actuated joints via fiber embedding in a polyjet 3D printing process. In: In *22nd Annual International Solid Freeform Fabrication Symposium* (2011)
20. "Neutron Imaging Setup," *Paul Scherrer Institut*. [Online]. Available: https://www.psi.ch/en/niag/neutron-imaging-setup

Dr. Amy Elliott is a full research staff at Oak Ridge National Laboratory where she serves as a principal investigator for binder jet additive manufacturing (AM). She and her team research technologies for a variety of industrial applications, including but not limited to new binders, densification methods for powder preforms made with AM, and new material systems which are compatible with indirect AM technologies. Dr. Elliott has received two R&D 100 awards, published over 50 journal articles, co-authored a book, and has several patents in the AM space. In 2012, Dr. Elliott was cast on Discovery Channel's "The Big Brain Theory," a reality show competition for engineers, where she placed second out of 10 contestants. In her free time, films as a science-personality for the Science Channel's Outrageous Acts of Science, explaining the engineering and science behind viral video clips. Dr. Elliott and her husband are also members of Eva Haakanson's electric racing team and have traveled to the salt flats in Utah and Australia pit crew for Eva's land-speed racing attempts.

Chapter 6
Science Personality and STEM Ambassador

Amy Elliott

Dr. Amy Elliott has had unique opportunities to serve as a STEM role model through her various media production projects. Her debut on screen was as a competitor on an engineering reality show competition called "The Big Brain Theory," aired by *The Discovery Channel*. Following this competition, Dr. Elliott served as an on-camera host for RoboNation's water-based, annual collegiate competitions and for *The Science Channel*'s viral video show, "Outrageous Acts of Science." Dr. Elliott also serves as a volunteer pit-crew for Green Envy Racing, a woman-lead race team that seeks to set and speed records with electric motorcycles. Because of Dr. Elliott's public exposure, she was selected as an IF/THEN Ambassador by the American Associate for the Advancement of Science (AAAS), a program that seeks to highlight women as STEM role models. Dr. Elliott's work in 3D printing is frequently a source of inspiration for her.

3D Printing Vending Machine

A decade ago, additive manufacturing was in a problematic position in that the technology was rising in popularity but was not very accessible, especially to the most interested party: students. The Virginia Tech DREAMS lab led by Dr. Chris Williams sought to remedy this situation by giving 24-hour access to 3D printers in the lobby of the Mechanical Engineering building through a device called the "DreamVendor." The DreamVendor was a kiosk that contained four Thing-O-Matic desktop-style 3D printers which the students could access through a keypad interface and SD card slot. The Thing-O-Matics were a unique style of 3D printer in that they included a conveyor belt configuration as the build platform, allowing the

A. Elliott (✉)
Oak Ridge National Laboratory, Knoxville, TN, USA
e-mail: elliottam@ornl.gov

© Springer Nature Switzerland AG 2021
S. M. DelVecchio (ed.), *Women in 3D Printing*, Women in Engineering and Science, https://doi.org/10.1007/978-3-030-70736-1_6

printer to eject the parts to the user. Dr. Amy Elliott (now at Oak Ridge National Lab) was the project lead for this endeavor as well as the lead for the design and assembly of the kiosk. The DreamVendor was deployed in the spring of 2012 and met with success [1]. The experimental nature of the project resulted in reliability issues related to the Thing-O-Matic conveyor belts, which would stretch and lose their texture over repeated use. This resulted in parts regularly delaminating from the build surface belt and in print failures. Newer versions of the DreamVendor have a solid build platform with a scraping mechanism. The patent for the original DreamVendor was issued in 2016 with Dr. Elliott listed as a co-inventor [3] (Fig. 6.1).

The Discovery Channel's "The Big Brain Theory: Pure Genius"

In addition to the DreamVendor project, Dr. Amy Elliott of Oak Ridge National Lab has had some unique opportunities as a STEM role model through many different media productions. During her graduate studies at Virginia Tech [4], Dr. Elliott was cast on a reality competition show by *The Discovery Channel* called "The Big Brain Theory," originally called "Top Engineer." The show was hosted by Actor Kal Penn, CEO of WET Design Mark Fuller, and Innovation Strategist Christine Gulbranson and focused on featuring engineers with hands-on skills creating innovative devices and machines in a competition environment [5]. The competition included 8 challenges, 10 contestants, and a famous guest judge for each challenge. The challenges included a waterfall-powered elevator, a mobile shelter for firefighters, and a food making machine to feed the masses, among other innovative concepts. The teams had to not only design solutions to these challenges but also build them in full scale with a fully stocked machine shop at their disposal. Each episode began with a unique problem and a "blueprint" challenge where the contestants would propose an innovative idea to solve the problem. Dr. Elliott was picked in the top 2 for 6 of the 8 blueprint challenges, landing her a captain's spot for all but 2 challenges. Dr. Elliott's teams won all but one challenge (the final challenge), where finalist Corey Fleishman and his team were able to build and deploy a bridge that held the weight of a small truck. The winning team chose a more practical approach to bridge building which led to success, while Dr. Elliott's team had chosen a more innovative approach (a self-assembling bridge) which proved to be too complex and heavy for deployment. Guest judges included NASA Astronaut Mike Massimino, Star Trek Visual Effects Supervisor Burt Dalton, Stock Car Driver Carl Edwards. Dr. Elliott placed second out of 10 contestants and got to meet Buzz Aldrin, who related to her runner-up placement at the end of the competition. The show received an Emmy nomination and the 2013 Science, Engineering, and Technology (SET) award by the Entertainment Industries Council [6] (Figs. 6.2 and 6.3).

Fig. 6.1 Original DreamVendor with Dr. Chris Williams and Dr. Amy Elliott,as a PhD student [1]

Fig. 6.2 Season Banner for "The Big Brain Theory: Pure Genius," Starring actor Kal Penn [2]

Fig. 6.3 Dr. Elliott inside a water column being constructed for the waterfall elevator challenge(left) and welding as part of another challenge (right) [2]

The Science Channel's Outrageous Acts of Science

Dr. Elliott's debut on "The Big Brain Theory" led to her role as a science commentator on a science show called "Outrageous Acts of Science," which airs on *The Science Channel* [7]. This show features viral science-related videos and the science behind them. For this show, Dr. Elliott studies the physics and science concepts behind each clip and explains them to the audience in a reachable way, using humor whenever possible. Dr. Elliott filmed 57 episodes for the show before the series was completed in 2018 (Fig. 6.4).

STRAPPING YOURSELF TO
CONSTRUCTION EQUIPMENT,

THIS CAN ONLY GO BADLY.

Fig. 6.4 Dr. Elliott on outrageous acts of science [7]

Dr. Elliott has also served as co-host for a series of educational robotics competitions including SeaPerch, RoboSub, and RobotX, which are sponsored by the Association for Unmanned Vehicle Systems International (AUVSI) Foundation. The competitions take place at numerous locations across the globe including Hawaii, Singapore, and San Diego and involve middle schoolers, high schoolers, and college competitors. The competitions focus on building water-based robots that can maneuver an obstacle course either through direct control by an operator (middle-school level) or autonomously (collegiate-level competitions). Unique challenges to the competition are the need to seal and waterproof all electronics, ensuring neutral buoyancy and balance of the water craft, and tuning of sensors for the water environment. For the collegiate competitions where autonomy was required, students battled with ever-changing lighting conditions above and below the water. Advanced techniques are demonstrated at the competitions, which Dr. Elliott could explore as she interviewed the teams during the competition.

Pit Crew for the Fastest Woman on a Motorcycle

Dr. Elliott and her husband share a love of motorsports and by chance were given the opportunity to serve as pit crew members for Dr. Eva Hakansson, who has dedicated her life to electric racing. Dr. Hakansson builds and tests her own electric, streamline motorcycles as a form of eco-activism, believing that by setting world land-speed records will show the advantages of electric drivetrains and speed their adoption. Thus far, Dr. Hakansson has set over 15 world records (including a Guinness World Record) and won herself the title of "Fastest Woman on a Motorcycle." In 2018 Eva moved from Denver, Colorado to New Zealand so that she can be closer to Lake Gairdner, Australia, which is one of the longest stretches of salt flats in the world [9]. Dr. Elliott and her husband Sam joined Eva's team on the salt flats in Bonneville, UTAH for Speedweek in 2017 and on the salt flats of Lake Gairdner Australia in 2019. Dr. Hakansson is also a 3D printing enthusiast,

Fig. 6.5 Dr. Amy Elliott and Dr. Zoz Brooks Hosting the 2014 Robosub Competition at the US Navy TRANSDEC Facility in San Diego, CA [8]

Fig. 6.6 Dr. Eva Hakansson (Center) and Dr. Amy Elliott (Left) with the Green Envy Race Team at the 2019 Speed Week on Lake Gairdner, Australia

printing many components of her streamliner using a home-made large-format 3D printer (Figs. 6.5 and 6.6).

American Association for the Advancement of Science IF/THEN Ambassador

In 2019 Dr. Elliott was selected as one of over a hundred women in STEM careers to be a American Association for the Advancement of Science (AAAS) IF/THEN Ambassador, whose namesake comes from the idea, "We truly believe that IF she can see it, THEN she can be it." As an IF/THEN Ambassador, Dr. Elliott has a public profile that is available to educators to access as well as a life-size statue on display at NorthPark Center in Dallas, Texas, USA, to serve as a recognition of women in STEM fields [10]. The ambassadorship is sponsored by AAAS foundation and Lyda Hill Philanthropies. During the 2-year term, each ambassador receives social media coaching to increase their impact and visibility as a woman in STEM. Specific focus is given to social media platforms like Instagram that are known to reach middle school girls (Fig. 6.7).

Fig. 6.7 Dr. Elliott's profile features materials for educators and students to learn more about women in STEM careers (left) and a life-sized statue that will go on display in 2021 [10]

References

1. Maly, T.: "The Future of Stuff: Vending Machine That Prints in 3-D." WIRED, May 2012. https://www.wired.com/2012/05/3-d-vending-machine/
2. "The Big Brain Theory: Pure Genius," Vudu.com. [Online]. Available: https://www.vudu.com/content/movies/details/The-Big-Brain-Theory-Pure-Genius-TV-Series-/444495
3. Williams, C.B., Elliott, A.M., McCarthy, D.L., Meisel, N.A.: 3D Printing Vending Machine. US20140288699 A1, 2014
4. Mackay, S.: Engineering doctoral student to compete on Discovery Channel's 'Big Brain Theory', VTNews.VT.edu, Blacksburg, VA, Apr-2013
5. The Big Brain Theory. 2013
6. "Star Trek Into Darkness, World War Z, Big Bang Theory, Mythbusters, Iron Man 3, Grey's Anatomy among Honorees for Annual Science, Engineering & Technology (SET) Awards," prweb.com, 2013. [Online]. Available: http://www.prweb.com/releases/2013/10/prweb11185202.htm
7. *Outrageous Acts of Science. Science Channel. https://www.sciencechannel.com/tv-shows/outrageous-acts-of-science/*
8. "RoboSub," robonation.org. [Online]. Available: https://robonation.org/programs/robosub/
9. "Lake Gairdner," *Britannica*. Encyclopædia Britannica, 2010
10. "AAAS IF/Then Collection – Amy Elliott," ifthencollection.org, 2020. [Online]. Available: https://www.ifthencollection.org/amy

Dr. Amy Elliott is a full research staff at Oak Ridge National Laboratory where she serves as a principal investigator for binder jet additive manufacturing (AM). She and her team research technologies for a variety of industrial applications, including but not limited to new binders, densification methods for powder preforms made with AM, and new material systems which are compatible with indirect AM technologies. Dr. Elliott has received two R&D 100 awards, published over 50 journal articles, co-authored a book, and has several patents in the AM space. In 2012, Dr. Elliott was cast on Discovery Channel's "The Big Brain Theory," a reality show competition for engineers, where she placed second out of 10 contestants. In her free time, films as a science-personality for the Science Channel's Outrageous Acts of Science, explaining the engineering and science behind viral video clips. Dr. Elliott and her husband are also members of Eva Haakanson's electric racing team and have traveled to the salt flats in Utah and Australia pit crew for Eva's land-speed racing attempts.

Chapter 7
Go Big or Go Home – Printing Concrete Buildings

Megan A. Kreiger

Automating Construction: Age-Old Industry Conversion to the Digital Realm

Construction and Additive Manufacturing

When people think of modernization, the first industry at the top of everyone's minds is often not construction. Construction is an industry that has been slow to change and adopt new technologies for reasons spanning from liability to practicality and cost effectiveness. This leaves the industry ripe for improvement that can be met by the incorporation of new technologies.

Additive manufacturing is an area that has seen rapid growth and innovation in the last 10 years. A decade ago, this industry largely consisted of a few large industry players, some wild concepts, and the "RepRap" hobbyists. The RepRap (self-REPlicating RAPid prototyper) movement [1] of makers, hobbyists, tinkerers, and professionals all around the globe led to an agile and ever-changing field, with iterations built on iterations of designs and printers themselves. This worldwide effort has evolved into a wide breadth of everyday 3D printers and innovations for many applications in a manufacturing environment whether at home or in a factory setting, as is discussed throughout this book.

M. A. Kreiger (✉)
U.S. Army Engineer Research and Development Center, Champaign, IL, USA
e-mail: erdcpao@usace.army.mil

© Springer Nature Switzerland AG 2021
S. M. DelVecchio (ed.), *Women in 3D Printing*, Women in Engineering and Science, https://doi.org/10.1007/978-3-030-70736-1_7

Step Toward Automation in Construction – Opposites Attract

The combination of these two seemingly opposite-styled fields, construction and additive manufacturing, into the subject area of additive construction (AC) provides an avenue for construction to take advantage of the rapid development that has persisted in the field of additive manufacturing over the last decade.

It is well established in the field of robotics that automation is intended for tasks that are dull, dirty, and dangerous. These three Ds of robotics describes many of the occupations within the construction industry. According to the US Occupational Safety and Health Administration (OSHA), 20% of all job-related fatalities in the United States are associated with construction [2]. Now imagine a more dangerous scenario where these tasks are performed while in a military environment. Automation in private sector and military construction will be essential to drive industry modernization, protect lives, reduce hard labor and work-related injuries, and improve efficiency.

The union of these areas is much like you would expect a marriage of opposites to be. Each field provides just what the other needs to push innovation toward a practical application. The intersection of additive manufacturing and construction is not for the light of heart; it is both innovative and fast paced which often leads to varying degrees of contention. As one would expect, the pace that additive manufacturing operates is frustrating for the construction industry. This results in a high amount of criticism and disbelief, which is enflamed by the "hype" released by the additive construction field. Likewise, the continuous pushback from the construction industry to slowdown and wait to allow building codes and testing to catchup is frustrating for those of the additive manufacturing mentality, so the two industries squabble over who should take the lead. The answer lies somewhere in-between; drive for building codes to be established while working to continuously improve the technology and expand applications. The perfect storm between push-pull is a key driver for innovation for the field of Additive Construction, and subsequently construction in general.

Think Big: The Adaptation of Additive Manufacturing to Additive Construction

From Lab-Top to Slab-Top

Coming out of the early 2010s, the additive manufacturing industry mostly consisted of 3D printers that were less than 1 ft^3 and came in two general varieties: (1) Expensive with decent quality meant for prototyping and (2) Less than $500 but requiring constant tweaking. The "RepRap" movement drove these inexpensive lab-top systems through constant improvements by sharing ideas and innovations via the open-source community, leading to today's widespread 3D printing capabilities.

During this time, Megan Kreiger, who would later work in the development of the field of Additive Construction, attended Michigan Technological University for a master's degree in materials science and engineering. Here her interests in 3D printing piqued through interactions with Dr. Joshua Pearce as part of the Michigan Tech Open-source Sustainability Technology (MOST) Laboratory. Within the MOST Laboratory, Ms. Kreiger engulfed herself into working with and maintaining a number of RepRaps within the laboratory, and applying 3D printing to ideas beyond prototypes. These efforts resulted in a thesis on "The use of life cycle analysis to reduce the environmental impact of materials in manufacturing," which encompassed recycling in a manufacturing environment, distributed recycling for reuse in 3D printers, and distributed manufacturing using 3D printers for end-use parts [3–5]. This work showcased that from a life cycle perspective, distributed manufacturing and recycling can reduce greenhouse gas emissions, carbon emissions, cost, and logistics compared to traditional methods.

While Megan was studying at Michigan Tech, there were a few exceptions to small-scale printers, notably the work at Loughborough University and the work by Behrokh Khoshnevis at University of Southern California, who today is sometimes referred to as the "Father of Large-scale 3D printing" [6]. These printers were mostly prototypes, one-offs, and only lived within research laboratory environments. As we inched toward 2015, there were a handful of printers around the world that were capable of construction scale with the majority focused on polymeric materials and not on traditional construction materials.

In 2015, the US Army Engineer Research and Development Center, Construction Engineering Research Laboratory (ERDC-CERL) initiated a research program on Automated Construction of Expeditionary Structures (ACES). The goal of this program was to print custom-designed infrastructure, on-demand, in the field, using locally available materials. Having graduated from Michigan Tech, Ms. Kreiger was hired as the subject matter expert on 3D printing for this research program. Her primary focus was to develop additive manufacturing technologies for additive construction, scaling a technology that has a typical print envelop of <1 m^3 to one of 50 m^3 or more. Her application was specifically for the military, with less emphasis on lab-scale or lab-top systems for controlled environments and instead, a heightened focus on deployability and ruggedization for outdoor in-situ printing directly on the ground or concrete slab top.

Challenges

The primary focus of the ACES project was to develop an additive construction system able to handle deployable operations. With this goal in mind, Megan Kreiger worked with the team to address multiple challenges related to additive construction, investigating specifically those that apply to concrete additive construction (i.e., concrete 3D printing) and the military. These challenges included: scalability, deployability, ease of operation, ease of repair, use of common easy to find parts, utilization of locally available materials, use of large aggregate, integration of traditional reinforcement and anchoring methods and materials, printable materials, and material flow systems. The majority of these challenges focused heavily on providing a new capability while reducing logistics for military construction, requiring an entirely unique approach with a system tightly constrained to the reduction or at most retention of existing logistics. In order to be applicable to the military, the 3D printer itself would need to be able to be repaired anywhere in the world, without burdening supply chain logistics. This means parts are able to be acquired local to the operation or that supply outside of the local area could not hinder operations. For this use case, concrete extrusion was primarily looked at as the methodology, as other methods are not as applicable in field conditions. The challenge that would prove to be king of them all was none other than one of the oldest: concrete.

Concrete

A mixture of cement, sand, rocks, and water; traditional concrete is the oldest particle-reinforced ceramic composite. Second to water in global use, concrete is the most popular material in the construction industry [7, 8]. It is utilized for its global availability, moldability, versatility, high compressive strength, and durability. Freshly mixed concrete can be delivered to a construction site in large quantities, can be molded to the shape of any form in which it is placed, and will harden under a curing chemical reaction at ambient conditions. Furthermore, concrete can be easily modified, or engineered, for many use cases by simply changing the proportions of the base constituents, and by adding "additives." Additives can come in the form of liquids, powders, or fibers. Additives can change many characteristics, including the curing rate of the cement-water reaction, the fresh state rheology (flow), the durability, and the cracking resistance. Unlike more traditional concretes, printable concrete materials have a number of unique requirements including the need to flow while in a pump (pumpable), but also not deform under its own weight once placed (shape stability).

When the ACES project began, the handful of research institutions exploring concrete additive construction were focused exclusively on mortar materials. Even today, most institutions and companies are 3D printing with mortars. Unlike concrete, mortars do not have large aggregates (≥ 9.5 mm rocks) and behave similar to a homogenous material akin to other 3D printing materials. Concretes, with the presence of larger aggregates, on the other hand, behave heterogenous in nature. This results in material behaviors that are harder to predict. Mortars also typically have a higher cement content, which increases their cost resulting in an inherant cost benefit for incorporating large aggregates. Under deployed conditions, the acquisition of concrete using locally available materials is preferred, as it represents the most likely scenario in the field. Similarly, reducing the amount of cement and cost is critical for production and logistics. Going into the ACES project, Megan and the team understood that printing with concrete was going to be a challenge, as the methods were not well documented in literature, and they would need to build this capability from the ground up.

From Concept to Reality, Pushing Against the Status Quo

Letting Go of Small-Scale Ideology

It is easy to think that you should "go with what you know" when designing a 3D printer that does a similar task to small scale. In this sense, many large-scale 3D printers follow the same design as their smaller predecessors, whether you are looking at gantry-style or delta-style printers. There is an inherent flaw with this mentality as there tends to be a viewpoint that you should design a scale model, prove that it works, and then build a bigger version of it. The majority of 3D printers rely heavily on precision rails and drive systems; one example of this is linear motion smooth rods or threaded rods. When scaling this component, the first shocking portion of this is the cost. While these components are relatively affordable when making a printer around 300 mm × 300 mm × 300 mm, when you scale the printer to 6 m × 12 m × 3 m, these costs end up being a substantial portion of the overall cost of the equipment. Similarly, mounting these components for the lengths required, while maintaining the rigidity required to prevent binding issues at this scale, provides an equally impractical issue. Instead of viewing the problem this way, Megan made the determination that scaling had to be the central design point for the 3D printer itself. Aside from the widespread issues surrounding the idea of scaling a 3D printing system, Megan had an application that needed a much tighter window of what would be acceptable – something that could be used for field applications for the military – deployability. None of the existing solutions or concepts pointed toward a deployable solution. The use of precision components in any form for this application were impractical for this problem set, as not only would they deteriorate, but when replacement was necessary, lengthy lead times for operational environments would result.

Easy-to-acquire components and parts that were simple to scale became the focus of Megan Kreiger's team. This led to an initial 3D printer design she built with her coworkers, a gantry system with an XYZ coordinate frame built out of scrap materials and a gantry crane, dubbed ACES 1 (Figs. 7.1 and 7.2). Utilizing an existing small gantry crane as the frame of the 3D printer provided the stiffness required and the scalability to larger gantry cranes readily available on the market. This design was unlike others in the field as it cast aside the use of precision rails and operated with a more flexible drive system, but after testing through multiple prints, necessary design changes became apparent. One of the major areas to change was the bridge for Y axis and Z axis (up/down). The use of the typical bridge of a gantry

Fig. 7.1 Megan Kreiger working on the initial prototype printer ACES 1, in 2015

Fig. 7.2 ACES 1 prototype printer with students Jacob Wagner and Jason Galtieri (left) and ACES 2 prototype with student Jacob Wagner (right)

printer proved dangerous for printing operations and highly limited the types and number of prints within a build envelope. Often during a print, personnel needed access to the print for various reasons including reinforcement placement and human intervention. These limitations required a full redesign to provide more access to the print bed, improve safety, and allow individual components to be printed at different heights and times. The next printer was designed with a fixed bridge as the Y axis and a vertical truss member as the Z axis to independently move up and down from the bridge. The design allowed for a more flexible build area without the risk of the bridge coming into contact with the printed structures. From here, Megan Kreiger proposed an innovative idea of utilizing an extendable rail system for the printer to drive on for the X axis. The long axis of the system did not require any precision rails as it was composed of large-scale commonly found rails such as I-beams or train rails connected end to end to achieve a variety of print lengths. Megan worked with two students, Jacob Wagner and Jason Galtieri, to flesh out this idea into a full concept 3D printer and with Bruce Macallister and Russ Northrup to bring to life the full-scale printer known as ACES 2 (Fig. 7.2).

After the establishment of a Cooperative Research and Development Agreement (CRADA) with Caterpillar Inc., these printers took on a new form with lessons learned and requirements developed by Megan and team. Underneath the CRADA, a mobile weather resistant printer was developed dubbed ACES Lite for its ability to be light and transportable. The development of this printer allowed Megan and the team to conduct demonstrations in distant locations. In April 2018, Megan led the team in the participation in the Maneuver Support Sustainment Protection Integration eXperiment at Fort Leonard Wood, Missouri, USA. The team pushed the technology to operate with materials locally available in the area for 3 weeks without cover, in weather ranging from warm days (~30 °C) to snowy days close to 0 °C. Megan and Brandy Diggs-McGee were able to establish a training plan and documentation, successfully training military personnel within one day to "train the trainer", for training the subsequent military team the next day (Fig. 7.3).

Go with the Flow – Rethinking Concrete

Traditional cast reinforced concrete construction consists of the assembly of form-work, placement of reinforcement, and pouring of concrete. The first two of these tasks are the most time consuming and the most expensive in this process due to being labor intensive. With the use of additive construction, the placement of form-work is eliminated, resulting in up to a 40% reduction, according to work done by Megan and the team. The huge cost reduction makes this technology highly appealing from a cost perspective [9].

Fig. 7.3 ACES Lite, the deployable printer, operating outdoors

Now that the concrete was being printed and not cast, the placement of reinforcement had to be re-envisioned. For concrete structures, reinforcement is required to improve strength under tensile forces and to prevent cracking due to volume changes. Typical reinforcing in cast concrete consists of continuous interconnected cages of steel bars, which can be difficult in some 3D printed structures. Rather than viewing this as a hindrance, the ACES team developed methods for placing reinforcement in printed structures by utilizing the conventional methods and materials used for reinforcement and connections in masonry and precast concrete construction [9]. Megan emphasized and encouraged this methodology to allow for an easy conversion to building codes and for obtaining materials in the field.

A long-standing benefit of concrete is that it can fill any form and take any shape at ambient conditions, and cure to create a material that is strong and durable. Traditional cast concrete utilizes formwork to provide support for the material while it cures enough to support its own weight and eventually the in-service loads (loads that the element supports during its service life). However, the production of complex geometrical shapes, even those that improve structural performance, are often cost prohibitive with formwork that is difficult and cumbersome to construct. Through the elimination of formwork by 3D printing, the design could now move outside the box to produce more complex geometries without the added cost of formwork. However, without formwork, the 3D printable concrete must be shape stable or hold its shape under its own weight.

The difficulty of this material is increased by the need for a printable concrete to flow consistently and effortlessly through a pump and hose used to deliver the material to the print nozzle. The requirements for the material to be able to flow through a pump hose and hold its shape after placement forced the team to rethink the concrete formulation to achieve these properties. This required a focus on the rheology, or flow characteristics, of the concrete. Megan and the team explored concrete material technologies that utilize rheology, such as self-consolidating concrete (SCC), curbing, slip forming, and shotcrete. This material was not created overnight and required extensive laboratory and print testing before landing on mixes that would work. During early laboratory testing of the equipment with the materials selected for printing, there were many issues related to consistency and material quality. These issues, while extremely frustrating to overcome, gave the team a wide-base of knowledge of what makes a concrete printable and how to deal with quality control issues on the fly, something that is essential to field operations [10] (Fig. 7.4).

Beyond the Laboratory, a Drive Toward Field Demonstrations

Fail Early, Fail Often, Fail Better – Test the Bounds

Without failure, one cannot learn how to troubleshoot. During laboratory tests and demonstrations, Ms. Kreiger didn't have a fear of failure, because if the team never tested the bounds of what the technology could do, they would never know their limits. This was especially true when looking at how high they could stack layers of

Fig. 7.4 Raw material quality field issues: Megan holding contaminant aggregates on the order of 12–25 mm aggregate (left); large aggregates found in sand (middle); inconsistent gradation from batch to batch (right)

concrete within a certain amount of time. It is best to have failure on smaller components, to know the operational limits for a successful build before scaling up to more critical full-scale structures. Early printing and testing of large components or structures allowed for many issues to become apparent that simply would go unnoticed until full-scale operations. This method improved the process with each following print. The parameters adjusted were similar to those seen with existing polymer printers but resulted in substantially different issues and solutions. The process parameters included aspects, such as print speeds, overhangs, nozzle paths, start/stop location, layer transitions, and transitions between components. Something as simple as where they started and stopped their prints could manipulate how much material was placed at these areas and cause the entire print to fail from excessive gaps or weight. To print designs like buildings and bridges, these aspects needed to be understood before the print of a semi-permanent or permanent structure. Megan and the team developed a printing methodology based on the characteristics of printed concrete on full wall-sections and small structures to reduce risk of failure, injury, and equipment issues on larger structures. Printing for long durations required significant changes for print operations and required the bounds to be pushed to predict future issues and address them.

Thriving on the Unpredictable

When printing or demonstrating in the field, the only thing you can count on is Murphy's law: anything that can go wrong will go wrong. Many people in this position would protect themselves from this idea by establishing controlled settings, controlled materials, and complete process control. In doing so, however, you lend yourself to only being able to operate within this controlled framework. Instead of doing this, Megan pushed the team to lean in and embrace the unpredictable nature of field demonstrations. The reason is simple: if you embrace it, you can plan for it and establish ways to troubleshoot, allowing the technology to operate in more harsh and unstable environments. Embracing this unpredictable nature wasn't an option for the technology Megan was developing; it was a requirement. If errors and inconsistencies happened in the lab or in the field with a team of scientists and engineers, the same issues were bound to be more prevalent when released into the field to be operated by military personnel. To do justice for the technology and to provide value to the warfighter, the technology developed by Megan and the team embraced the idea of a wide breadth of site and materials conditions to operate in.

The ability to handle site and material condition variability was pushed forward in multiple demonstrations by Megan and team, resulting in the iterative improvement of the technology. The team initially started out testing the equipment in Champaign, Illinois, USA, in a controlled laboratory environment, then moved to a non-climate controlled uninsulated tent structure, followed by moving outdoors. This testing resulted in two 47.5 m^2 buildings (i.e., barracks huts) (Fig. 7.5), multiple smaller scaled structures, and traffic barriers. After the first building, the leadership Megan provided as the Technical Lead eventually led to her role as the program manager of the ERDC-CERL Additive Construction program.

Next, the team left their home base which brought a whole new level of challenges. Megan led the team to assemble an "away kit" to allow for the ability to print at Fort Leonard Wood, Missouri, USA, in all-weather conditions in a remote operational environment. This was only the second AC demonstration with military personnel, the first being the printing of test walls, led by Ms. Kreiger 8 months earlier. Planning for the demonstration included conducting a safety assessment, developing safety plans, and establishing training protocols. During this demonstration, Ms. Kreiger and the team were able to train one group of Army engineers within 8 h. This group of engineers then trained the next group of Army engineers without intervention. Furthermore, this demonstration highlighted the technologies capability to be deployed, use locally available materials, and continuously produce infrastructure components (e.g., t-walls and culverts).

Fig. 7.5 First full-scale 3D printed building in the Americas with ACES 2 (left), enhanced building printed by ACES Lite (right)

The demonstration at Fort Leonard Wood was followed by the printing of a structurally enhanced building structure just 5 months later. This demonstration was intended to test continuous day-and-night operations and test the "24-h building" claim that was prevalent in the field. The demonstration allowed the team to determine that it was possible to complete a structure of a 46.5 m^2 building within 48 h. If equipment improvements are made, printing the walls within 24 h would be possible [11]. In order to align with the industry and comply with current codes and standards, Ms. Kreiger and the team collaborated with Skidmore, Owings, and Merrill (SOM), an architectural/engineering firm in Chicago, Illinois. Work on the barracks hut (a.k.a. b-hut) structure further developed the team's reinforcing strategies and the use of structural optimization. The walls were designed to be self-stable with a triangular wave pattern at the foundation that morphed to a straight pattern near the roof. This "chevron" design makes it possible to produce different configurations, lengths, and heights, with minimal changes to reinforcement.

Within a mere 3 months' time after the enhanced b-hut print and notice, Megan and the team demonstrated the ability to move quickly from design to construction, by performing a complete design and print of the first 3D printed bridge in the Americas (Fig. 7.6). The bridge was printed near site and construction at Camp Pendleton, California, USA. The bridge was printed and emplaced in a military training dry gap within 6 days. The bridge consisted of three spans and two piers. After printing the first bridge in the Americas, the bridge was shipped to Champaign, Illinois, and so members of the ERDC AC team could perform destructive testing. The team intentionally broke the printed bridge, documenting that it could support 20,400 kg, which was three times the design requirement of 6800 kg.

From Novelty to Adoption – Looking Forward

Once a series of one-offs and experimental printers in the early 2010s, additive construction is now a blossoming field of study and promising construction practice. However, this unique construction method requires additional efforts before adoption in the construction industry. The technology must demonstrate acceptable construction and safety practices, design procedures, structural and material performance, and established codes and standards. For successful integration of this technology into the military, it must be tested in a full operational environment.

Megan Kreiger and the ERDC Additive Construction team (Fig 7.7) continue to address these items and drive this field forward through their research and operational demonstrations [12, 13]. To promote adoption, Megan and the team work with the industry to integrate additive construction into construction practices and standards. The ERDC AC team is actively engaged in the American Concrete Institute (ACI), American Society of Civil Engineers (ASCE), and American Society of Testing and Materials (ASTM). They are members of the planning and scientific committee for the 2019 and 2021 Transportation Research Board's (TRB's) International Conference on 3D Printing and Transportation and judges for

Fig. 7.6 First 3D printed bridge in the Americas

Fig. 7.7 Megan Kreiger – Program Manager & Subject Matter Expert (bottom left), Brandy Diggs-McGee – Operational Lead (bottom middle) & the ERDC-CERL Additive Construction team in front of the enhanced 3D printed building (2019)

the NASA 3D Printed Habitat Challenge. Megan and members of the team have presented at conferences hosted by the RILEM Digital Concrete, the North Carolina Transportation Summit, the National Science Foundation (NSF), ACI, ASTM, the National Ready Mix Concrete Association (NRMCA), TRB, and more. Megan actively reaches out to academia and other government laboratories. The efforts led by Megan Kreiger have been awarded by Engineer News Record, Frontiers of Engineering, BuiltWorlds', US Army Corps of Engineers (USACE), Michigan Tech, NASA, and ERDC. Similarly, these efforts have been featured by news agencies around the globe including Fox & Friends, 3dprint.com, USA Today, Engineering News Record, The Military Engineer, Popular Mechanics, ASCE Civil Engineer Journal, Engineer Your Career podcast, and more. Megan and the team continue working with equipment manufacturers, material developers, and active-duty personnel to ensure that the technology being developed is applicable to the military needs.

While the path to adoption of new technologies is often arduous, the potential benefits of additive construction are key drivers in the fast pace and adoption of this technology. The freedom of design for element creation associated with additive technologies will provide architects and engineers across the general and military construction industries the ability to implement designs they were once told were impractical or too costly. This newfound freedom allows for advancements in areas such as aesthetics, energy efficiency, structural design, material design, structural performance, and design optimization. Contractors will have the opportunity to advance the automation of construction using printer systems and the ability to track print times and operational usage, while reducing injury and arduous labor for workers. The technology has potential to reduce the cost and time of construction ensuring more competitive construction projects. Additionally, the use of additive construction will attract prospective employees to the construction industry due to the evolution of skill sets from general to technical. Additive construction has the potential to transform the built world, for all of us, one layer at a time.

References

1. Jones, R., Haufe, P., Sells, E., Iravani, P., Olliver, V., Palmer, C., Bowyer, A.: RepRap – the replicating rapid prototyper. Robotica. **29**(1), 177–191 (2011). https://doi.org/10.1017/S026357471000069XRepRap.org
2. U.S. Department of Labor, Bureau of Labor Statistics, in cooperation with state, New York City, District of Columbia, and federal agencies. Census of fatal occupational injuries. U.S. Department of Labor, District of Columbia (2019)
3. Kreiger, M.A.: The use of life-cycle analysis to reduce the environmental impact of materials in manufacturing. Master's thesis, Michigan Technological University (2012)
4. Tymrak, B.M., Kreiger, M., Pearce, J.: Mechanical properties of components fabricated with open-source 3-D printers under realistic environmental conditions. Mater. Des. **58**, 242–246 (2014). https://doi.org/10.1016/j.matdes.2014.02.038
5. Kreiger, M.A., Mulder, M.L., Glover, A.G., Pearce, J.: Life cycle analysis of distributed recycling of post-consumer high density polyethylene for 3-D printing filament. J. Clean. Prod. **70**, 90–96 (2014). https://doi.org/10.1016/j.jclepro.2014.02.009
6. Contour Crafting Inventor Dr. Khoshnevis: Widespread 3D printed homes in 5 years, high-rises in 10 years. https://3dprint.com/53437/contour-crafting-dr-khoshnevis/ (Mar 31, 2015)
7. The Cement Sustainability Initiative: Recycling Concrete, pp. 1–42. World Business Council for Sustainable Development, Geneva/Washington, DC (2009)

8. Gagg, C.: Cement and concrete as an engineering material: an historic appraisal and case study analysis. Eng. Fail. Anal. **40**, 114–140 (2014). https://doi.org/10.1016/j.engfailanal.2014.02.004
9. Kreiger, E., Kreiger, M., Case, M.: Development of the construction processes for reinforced additively constructed concrete. Addit. Manuf. **28**, 39–49 (2019). https://doi.org/10.1016/j.addma.2019.02.015
10. Kreiger, E., Diggs-McGee, B., Wood, T., MacAllister, B., Kreiger, M.: Field considerations for deploying additive construction. In: Bos, F., Lucas, S., Wolfs, R., Salet, T. (eds.) Second RILEM International Conference on Concrete and Digital Fabrication. Springer, Cham (2020). https://doi.org/10.1007/978-3-030-49916-7_109
11. Diggs-McGee, B., Kreiger, E., Kreiger, M., Case, M.: Print time vs. elapsed time: a temporal analysis of a continuous printing operation for additive constructed concrete. Addit. Manuf. **28**, 205–214 (2019). https://doi.org/10.1016/j.addma.2019.04.008
12. Jagoda, J., Diggs-McGee, B., Kreiger, M., Schuldt, S.: The viability and simplicity of 3D-printed construction: a military case study. Inf. Dent. **5**, 35 (2020). https://doi.org/10.3390/infrastructures5040035
13. Jagoda, J., Case, M., Diggs-McGee, B., Kreiger, E., Kreiger, M., Schuldt, S.: The benefits and challenges of on-site 3D-printed construction: a case study. In: Proceedings of the 3rd International Conference on Engineering Technology and Innovation, pp. 21–29. CNR Group Publishing, Belgrade (2019)

Megan A. Kreiger is a research mechanical engineer at the US Army Engineer Research and Development Center (ERDC) – Construction Engineering Research Laboratory (CERL). She is a pioneer in the field of additive construction. As the program manager for the additive construction program at the ERDC-CERL, Ms. Kreiger has taken a new approach to the modernization of the construction industry through the scalability of 3D printing, developing deployable additive construction technology. She realized that current automated construction methods were not suitable for use in austere environments. Ms. Kreiger pushed the technology to its outer limits, leading to the development of large-scale deployable construction 3D printers that can print on varied terrain, promote mobility, ease of build, flexibility of materials, faster construction speeds, and are designed to be rugged and print out in the open.

She focuses on how to use the technology today, through integrating reinforcement strategies and materials (including large aggregate) similar to traditional construction practices, while prioritizing materials that can be found anywhere in the world. Without the use of formwork, this technology can create complex custom concrete structures to enhance aesthetic appeal and structural performance at lower cost, all while making concrete work less physically taxing. Ms. Kreiger has led multiple printing demonstrations, most of which are the first in the Americas if not the first of their kind in the world. These include:

512 ft² Barracks Hut, with vertically straight printed walls and wood roof – Summer 2017

Army Maneuver Support, Sustainment, Protection & Integration Experiment (MSSPIX), Infrastructure Components (Barriers, culverts, fighting position) – Spring 2018

Enhanced 512 ft^2 Barracks Hut, with sloped walls and concrete roof – Summer 2018

33 foot 3 span bridge with two piers – Winter 2018

Ms. Kreiger has been described as having a passion and vision unrivaled in the industry of large-scale 3D printing, which acts as kindling to ignite others' imaginations for how far this technology will take humanity in the future. She promotes honesty and integrity in her field and is not afraid to question the status quo in a field focused on achieving perfection and promoting hype. She is internationally recognized as a 2020 BuiltWorlds' Maverick, 2018 ENR Top 25 Newsmaker, 2018 National Academy of Engineers – Frontiers of Engineering participant, 2018 USACE Innovation Award recipient, 2021 Michigan Tech Outstanding Young Alumni, and 2021 NASA Silver Medal Achievement - Team award recipient.

Ms. Kreiger fell in love with 3D printing through her graduate studies. She earned her M.S. in material science and engineering and graduate certificate in sustainable futures from Michigan Technological University in 2012, after earning a bachelor of science in mathematics in 2009. Ms. Kreiger's thesis was on "The use of life cycle analysis to reduce the environmental impact of materials in manufacturing." She operated and maintained RepRap 3D printers in the Michigan Tech Open Sustainability Technology group (MOST group). After graduating, Ms. Kreiger began working at CERL as a post-graduate in 2013, then became hired on as the start of the Automated Construction of Expeditionary Structures (ACES) project in 2015.

Chapter 8
Entrepreneurship and Innovation in Metal Additive Manufacturing

Melanie A. Lang

Introduction: From Aerospace Engineer to Entrepreneur

I credit my fascination with space from an early age, my enjoyment of math and science, and affinity to the arts as the driving factor to becoming an engineer. My desire to pursue a career that would hone technical skills but allow my creative side to flourish led to Aerospace Engineering, and eventually to 3D printing.

As a co-founder of a startup, the job is successfully executing a little of everything. From helping develop the product roadmap and completing R&D projects, to coordinating communications, recruiting and fundraising, to acquiring new customers along with facilitating business relationships, I find myself drawing on, and thankful for, every role I've held over the prior 15+ years in the Aerospace and Defense Industry.

The journey of becoming a female engineer to entrepreneur has been an exciting and challenging journey, driven by passion and a hunger for adventure. The impact metal additive manufacturing (AM) technology is imparting on multiple industries to open the design space is incredible, including redefining what is possible with multiple materials, embedded sensors, and complex geometries. I look forward to sharing my personal story of the journey and technology developments, while it continues to unfold.

M. A. Lang (✉)
FormAlloy Technologies, Inc., Spring Valley, CA, USA

© Springer Nature Switzerland AG 2021
S. M. DelVecchio (ed.), *Women in 3D Printing*, Women in Engineering and
Science, https://doi.org/10.1007/978-3-030-70736-1_8

Lessons Learned as an Aerospace Engineer

During my youth, the idea of creativity without boundaries in the realm of science was appealing because of my love of both art and science. However, what I found out is that the limitations, bounds, and/or requirements levied on the engineer limit the creative aspect, i.e., put a box around the problem. The maintainability and supportability, often grouped with usability and other system features commonly referred to as "illities," of a system was a challenge and continues to be a challenge because it's not something that is taught in school or that adolescents are typically exposed to. And right or wrong, I viewed these necessary aspects as limitations. The manufacturability, reliability, and supportability of the system or component could often drive the entire design and cost, and as a creative person that was a source of frustration.

On a more personal note, one of the profound learnings is that gender plays a role in the opportunities and outcomes of your professional endeavors. More specifically, being a woman in engineering and technology has its detriments, but also advantages, as is the case with many aspects of life. I've had experiences dealing with people who assume that I am not intelligent or do not deserve my position all the way across the spectrum to people who hold me on a pedestal, based almost solely on gender. I have also learned that the most successful outcome for me has been to present the best version of myself, regardless of another person's gender (or other) bias. "When they go low, we go high," mentality.

I believe that gender inequality in pay and opportunities is a well-documented issue, some of which I have experienced personally. I also believe there are benefits that I find are less often discussed. For example, I have had incredible female supporters and mentors who were willing to take me under their wing and help me to avoid some of the struggles and pitfalls that they had experienced. My experience has been almost entirely positive with women helping other women. Several years ago, as a new player in the industrial 3D printing space, I was lucky enough to have learned about Women in 3D Printing (Wi3DP). This incredible organization centered itself around the idea that we should embrace those diverse backgrounds because it is they who make the additive manufacturing industry so rich and interesting. Joining Wi3DP in my early days in this space helped me connect with suppliers, customers, supporters, and people I am now proud to call dear friends. As an entrepreneur, I have found several opportunities for grants, incubators, and mentorships that have helped thrust me forward and would not have been possible without a female founder.

Naysayers, distractions, roadblocks, and disadvantages will be there, technically, financially, and personally. As a female, an engineer, a leader, and a person, I attribute success to staying focused on the opportunities that can be created based on personal uniqueness and passion. AM has created another opportunity for women to be innovators, technology leaders, and entrepreneurs.

The Why, What, and How

Having a successful career in the aerospace and defense industry, I was certainly comfortable, and I often get questions related to why I became an engineer, why I became an entrepreneur, what I did, and how I did it. I hope my story provides insight into the abstract, and often wild, path of an entrepreneur.

The WHY. To understand why a company was founded, it's often interesting to understand the roots from a founder's perspective. In my case, being awed by space, inspired by the hobbyist maker movement and challenged by existing limitations to manufacturing in the aerospace and defense sector, these factors caused the giant metaphorical light-bulb to not only go off, but strobe intensely. The merger of art and science is always where my brain gravitated since a young age. I loved, and still love, being creative but also understanding the science behind how and why. At that intersection, to me, is engineering…you can create within a set of science-based principles. My deep curiosity of outer space and model-rocket-launching hobby guided me to aerospace engineering.

Once I entered the aerospace industry, I got my first taste of 3D printing for an Engineer's Week project in the early 2000s. National Engineer's Week is celebrated in the United States each year in February by educational institutions, societies, and companies to celebrate and encourage learning in the science and technology space, and always a favorite week of mine. That particular year, the design challenge included 3D printing of our design with stereolithography (SLA). Watching a design come to fruition, being built up layer by layer on-demand, was incredible to see. I kept that wonder and awe in my back pocket for several years before I entered the space as a hobbyist and 3D printing enthusiast and helped to build my own polymer 3D printer. In my professional life, I was grappling with the realities of supply chain and sustainability. It was a challenge for me to accept that designs and innovation are limited not by the human mind or creativity but by manufacturing and supply chain. I was able to build what I wanted when I wanted on my home 3D printing system. I began to ask myself – what if an engineer could build what they wanted (out of metal) to make superior parts and systems? Enter the ideation state for FormAlloy, metal deposition technology for research, aerospace, energy, functionally graded materials, and future applications the next generation of designers come up with (Fig. 8.1).

The HOW. What began with a space-enthusiast/aerospace engineer and mechanical engineer building a small hobbyist printer on a kitchen table to where FormAlloy Technologies, Inc. and our team are today has been quite the ride. Once the driving force or problem statement for the company was defined, getting started was challenging, yet exciting. From my experience having more than one founder is a good thing, particularly if they are with you from the beginning. Startups require an unequivocal level of energy and focus, and having someone to strategize, criticize, and reinforce was critical. Just as important as a trustworthy, passionate founding team is to have solid business mentors. Joining incubators and business accelerators

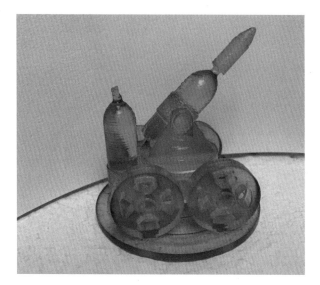

Fig. 8.1 First 3DPrinted prototype via sterolithography

expose you to finding those key mentors that will, in my case at least, be with you as trusted advisors for much longer than the program itself.

Once the foundation, as in the idea and the team, was set, it was time to start building, or in my case, expanding on what was already built, into a prototype. After the working prototype it was time for market and/or customer validation. For my journey, that was bringing the FormAlloy system to the RAPID + TCT event, touted as North America's largest and most important additive manufacturing event, in 2016. The market for the concept was validated, with significant interest and multiple initial contracts from NASA, large corporations, and small businesses alike. With the demand for research and development (R&D) projects and feasibility studies strong, it was time to grow the team and make a strategic decision to seek investment or bootstrap and continue to self-fund. With the significant early traction, we were able to self-fund and began to scale our team and production. As our demand increased, we required rapid growth and sought investment from a strategic partner and successfully closed a Series A funding round in 2019. And this cycle continued and continues today, technology advancement, revenue growth, team growth, and strategic decisions. Entrepreneurship in general and fundraising in particular is both a science and art, and I have found it fascinating.

WHAT. FormAlloy Technologies, Inc. is an original equipment manufacturer (OEM) for metal deposition technology used in research, aerospace, energy, functionally graded materials, and future applications the next generation of designers come up with. In addition to providing AM systems, FormAlloy provides OEM

solutions with production in mind, enabling customers to integrate the same components from the metal deposition systems into their existing manufacturing cells and
production lines. Although some customers use metal AM systems to develop their
own intellectual property (IP), other customers outsource production and even
R&D. For those customers, FormAlloy offers metal deposition services as well for
R&D projects, feasibility studies, and production applications.

Innovation in Metal Additive Manufacturing

The impact the metal additive manufacturing technology is having on multiple
industries to open the design space is incredible, including redefining what is possible with multiple materials, embedded sensors, and complex geometries. The
metal AM ecosystem is enabling innovation in everything from large aerospace
components, to consumer products, to undersea systems, and the potential is even
more fascinating. My focus with FormAlloy is to deliver a technology that can be
used to drive innovation and improve performance, with quality and repeatability.
However, before diving into innovation specific to metal deposition technology, I
want to acknowledge the wide range of technologies available for metal applications.

The Metal AM Ecosystem – From Micro to Massive

Similar to traditional manufacturing, there are many processes to select from within
the additive manufacturing technology umbrella. There is no "one size fits all."
Additive manufacturing systems and processes, along with traditional manufacturing processes, should be considered tools in the toolbox. To understand the range of
technologies available in metal AM, an ecosystem of process types and the scale
they represent are useful. With a growing number of metal AM processes and companies coming available, the ecosystem is ever-changing and growing. This snap-
shot in time of the ecosystem, as seen in Fig. 8.2, can be useful to gain a basic
understanding of the diversity of systems and solutions available. I encourage new
technology seekers to investigate each technology and talk to a range of suppliers
and experts to choose which technology or technologies are a best fit for your unique
set of applications.

Technologies in additive manufacturing provide a range of capability from small-
scale to large-scale, with a variety of build, material, speed, and quality capabilities
[1]. Some of these technologies are used for micro-sized components, while others
including metal deposition technology are scalable for large components. And due
to the rapid pace of technology development in this space, new printers and new
innovations are always being released.

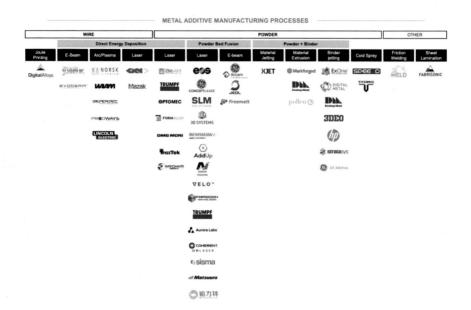

Fig. 8.2 Metal AM ecosystem. (Courtesy of Digital Alloys [1])

Application-Driven Technology Selection

Now that a wide range of technologies exist, with even more in the pipeline, implementation is based on a requirements-driven approach for technology selection. I recommend a weighted or prioritized matrix that consists of the following elements, tailored to the application(s):

- Materials that will be used for printing
- Materials that will be used as substrates, if required
- Resulting surface finish
- Processing speed
- Part resolution

Adopting a weighted decision matrix based on the priorities for a particular application is critical for success in additive manufacturing, to ensure your innovation comes to fruition.

Metal Deposition – The Triathlete of AM

Because of the diverse range of technologies and potential with metal additive manufacturing, I will not claim one particular technology is better than other; that is for the user to determine. I *will* claim that metal deposition is the most interesting and profound to me and my FormAlloy journey. The fascinating aspect of metal deposition is that it can be compared to a triathlete in the metal additive manufacturing space. FormAlloy's metal deposition technology utilizes a directed energy deposition (DED) process, also referred to as laser metal deposition (LMD). The technology enables its users to handle several different applications including building new components, enhancing components and repairing components for wear, manufacturing errors, and/or design changes. The ability to add value to components and builds in a variety of ways provides an open slate for innovation and technology advancement. At FormAlloy, we market the ability to *FORM, ENHANCE,* and *REPAIR*, which covers a broad range of potential for metal deposition. The innovations required to deliver the quality, repeatability, and component performance demanded by high-value applications are non-trivial and require significant skill, technology, and creativity to achieve. Systems that provide build rates of 2–100× faster, powder efficiency that exceeds 90%, and in-situ monitoring and control to ensure quality, are some of the aspects that enable adoption of this particular additive manufacturing method. In order to reduce parameter development time and certify production builds, a combination of real-time process monitoring and control, combined with build inspection and data collection, summarize the innovative developments FormAlloy has implemented. In addition to processing aspects, the ability to achieve builds that were not previously possible, such as functionally gradient material (FGM) systems to give superior performance properties, is the real motivation to keep pushing ahead.

Exotic, Functionally Graded Materials, and Alloy Development – Re-defining What Is Possible

I am personally driven by re-defining what is possible, by using metal AM for a wide range of materials, geometries, and applications. Metal deposition is a well-suited technology for the formation of Functionally Graded Materials (FGM) systems due to its ability to use multiple materials within the same build, open-source parameters, and low material consumption. However, combining dissimilar materials delivers another set of challenges in materials performance and bonding. In order to reduce some of the consequences of multi-material deposition, such as residual stresses between the materials, a gradient transition can be used. During a gradient build, alloy compositions can be varied slowly, rather than simply adding one material directly to another. As always, the preferred method depends on the application and desired end properties. As more data becomes available for gradient transitions, benefits can be seen in many cases. In addition to the challenges inherent in multiple material builds, the ability to deposit the correct variations of materials at the correct time provides another hurdle to overcome. In the past, the process of multi-material builds relied upon on-the-fly blending or introducing additional powder feeders to deposit the transitions. But for intricate transitions, especially gradient strategies that may be generated by computational methods, this is not a viable long-term solution. In addition, the ability of the metal deposition technology to deposit multiple materials within the same build program could lend itself to discovering new alloy compositions. Pre-blending several compositional variants and depositing the potential next game-changing alloy is possible, but also with the same limitations as FGMs with several transitions. Limitations and challenges are frustrating, but can also be viewed as a catalyst for new ideas and innovations.

With both FGM and new alloy development use cases in mind, FormAlloy developed the Alloy Development Feeder (ADF). Within the ADF (Fig. 8.3), up to 16 different compositions can be deposited quickly. A revolver style hopper system enables rapid transition from one hopper to another, and feeder operation features help prevent cross-contamination between materials. Users now have a single feeder solution to produce the most complex FGMs imaginable and rapidly deposit new alloy after new alloy with ease. With our new feeder system in use by customers to develop FGM components and research new alloys that will improve additive manufacturing in general, it is beyond rewarding to see a technology delivered and making an impact that started as a problem [2]. On behalf of the FormAlloy team, I was honored to receive an innovation award from SME and provide a key note presentation during RAPID + TCT 2019 on our impacts to AM related to new alloy development (Fig. 8.4).

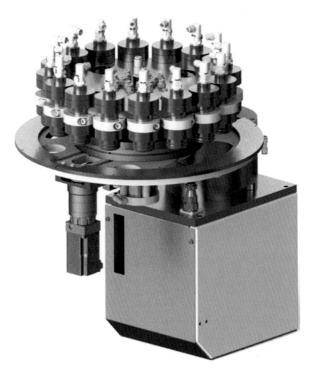

Fig. 8.3 Innovation in metal AM – the Alloy Development Feeder. (Photo courtesy of FormAlloy Technologies, Inc.)

Fig. 8.4 Innovation Award Recipient, Melanie Lang, with SME leaders at RAPID + TCT conference

Metal AM for Aerospace and Beyond

The aerospace industry was one of the earliest adopters and proponent of rapid growth for metal AM. From accelerated product development and complex geometries to testing and flight certification, aerospace has utilized the technology to deliver aircrafts with improved efficiency. The driving force for technology adoption within aerospace is access to space. 3D printing is enabling more companies to explore and utilize outer space, which would not be possible without the technology. Although aerospace provides an interesting use case for additive manufacturing, it is only one industry that makes up the AM ecosystem. Applications from rocket nozzles and oil and gas, to automotive tooling and consumer goods bring a diverse set of benefits and challenges to an even more diverse user base.

An Overview of Work in the Aerospace Sector

Although my focus is in metal deposition, metal AM components are produced from a variety of technologies. Currently, more widely adopted processes include laser powder bed fusion (LPBF), and electron beam powder bed fusion (EBPBF). The adoption of directed energy deposition continues to increase as the innovations are enabling the technology's ability to reach the necessary quality and repeatability demands. As previously discussed, each process has its advantages and disadvantages over the others in terms of producing structures and components. While LPBF has the highest resolution and best surface roughness as printed, the other technologies provide higher productivity, wider material processing, and more scalable capabilities. AM users within the aerospace sector know there is not a one-stop solution for their current and future manufacturing needs and continue to support the advancement of metal AM processes. Some of our most exciting projects in aerospace have focused on a hybrid use of multiple AM processes.

While metal deposition is increasing in adoption, one of the most public success cases using metal AM in aerospace comes from GE Aviation and their 3D printed LEAP engine fuel nozzle. The fuel nozzle produced via AM leads to part consolidation from 20 individual welded parts to a single printed piece, reduced part weight by 25%, and reduced lead times. The fuel nozzle is one of the contributing components to improve the LEAP engine efficiency by up to 15%. Since the development, GE has proceeded to print these fuel nozzles in production quantities to support engine orders and in 2018, the company celebrated their 30,000th printed fuel nozzle [3] (Fig. 8.5).

With NASA as one of drivers of metal AM and in the United States, rocket nozzles are another widely publicized use case that often utilizes metal deposition or DED technology. Faster design iterations are accelerating space development, but in addition to development, demands for larger rocket components are increasing [4]. Recent efforts are focused on increased build volume and accompanying

Fig. 8.5 GE Aviation's 3D printed LEAP engine fuel nozzle [3]

Fig. 8.6 Large rocket nozzles with complex geometries such as integral cooling channels built with DED. (Photo courtesy of NASA [4])

productivity, without detriment to the overall part quality. Large rocket nozzles with complex geometries such as integral cooling channels can be built as seen in Fig. 8.6.

As an aerospace engineer at heart, I am proud to say the NASA examples are some of the best use cases for large-scale AM such as DED. This process can form the integral channels for nozzles and combustion chambers, build bi-metallic components, and significantly cuts down on lead times and costs of traditional methods. Lead times can be reduced to weeks or months from years [5]. For this type of high-value low volume application, costs, including tooling costs, can be significantly reduced. With the on-going work in aerospace, I firmly believe we are not yet reaching the limit of what is possible. For example, 3D printing in space, building extra-terrestrial habitats, and embedding sensors into builds are within reach.

Trends and Emerging Challenges

Going forward, adoption rates for larger-scale process, such as metal deposition, will continue to mature similar to how the powder bed and binder jet processes have over the last decade. As distortion within large builds decrease, process software development improves for tool path generation, and closed loop control adaption continues, the increased quality and repeatability will enable customers to expand both in size and throughput. As the technology improves in feature resolution, build volumes and productivity increase, and multi-material components are designed, metal deposition technology will continue to deliver new solutions beyond the capabilities of traditional manufacturing processes.

The creativity and innovation of a new generation of designs that are not limited by traditional manufacturing methods or materials brings a new set of challenges. I believe the next wave of challenges related to AM will be on the materials side. Users will demand better performing materials that are optimized for a specific application, rather than relying on materials that have been used traditionally for decades or even centuries. I am proud to be a part of pushing the bounds of what is possible with metal deposition, helping additive manufacturing move forward as a whole, and providing systems and solutions that enables new materials, designs, and performance.

Innovation Never Stops

In order to push the bounds of what is possible and continue moving forward, company culture is key. Determining an approach and strategy to harness the incredible brain power of the collective FormAlloy team has been rewarding. Seeds for ideas are regularly generated by engaging employees across different functions with end users and customers through customer discovery. Customer discovery is understanding the real and perceived pain points of a user, either by "getting into the Gemba" (i.e., inserting yourself into the work environment to experience it first-hand) or by engaging with customers by asking questions regarding successes and challenges, and what they would change if they could change anything. In addition to the gratitude I have to my own team, I am personally grateful to the AM community at large for sharing experiences and challenges, which have paved the way to innovation.

Overall, executing an innovation-focused strategy has not happened entirely organically; it has been built into the metrics, discussions, rewards, and business rhythm of the company. If change is required to adopt a more innovative culture in your company, document where you are, where you want to go, and measure it along the way.

Acknowledgments The work reported here was a collaborative effort, and as such acknowledgement is due to others that contributed to the experimental work and/or critical discussions outside of those already called out in the text. This includes researchers at FormAlloy Technologies, Inc. (Mr. Jeff Riemann), Relativity Space (Mr. Sam Tonneslan), Trumpf (Dr. Eliana Fu), University of California San Diego (Dr. Kenneth Vecchio), and NASA Marshall Space Flight Center (Dr. Paul Gradl). The views expressed here are those of the author's and do not reflect the official policy or position of FormAlloy Technologies, Inc. or any of its customers, suppliers, and collaborators.

References

1. Application Criteria | Digital Alloys. Accessed Dec 22, 2020. https://www.digitalalloys.com/blog/application-criteria-metal-additive-manufacturing/
2. Dippoa, O.F., Kaufmannb, K.R., Vecchio, K.: High-Throughput Rapid Experimental Alloy Development (HT-READ). Published online 2020 in National Academies' National Materials and Manufacturing Board (NMMB)
3. 3D Printing Media Network – The Pulse of the AM Industry. GE Aviation already 3D printed 30,000 fuel nozzles for its LEAP engine. Accessed Dec 22, 2020. https://www.3dprintingmedia.network/ge-aviation-already-3d-printed-30000-fuel-nozzles-for-its-leap-engine/
4. Gradl, P.R., Protz, C., Wammen, T.: Additive manufacturing development and hot-fire testing directed energy deposition inconel 625 and JBK-75 alloys. In: 55th AIAA/SAE/ASEE Joint Propulsion Conference, pp. 1–20 (2019)
5. Gradl, P., Protz, C., Fikes, J., Ellis, D., Evans, L.: Large Scale and Multi-Alloy Rocket Engine Component Development using Various Metal Additive Manufacturing Techniques. Published online 2020

Melanie A. Lang, FormAlloy Technologies, Inc. co-founder and CEO, is motivated by developing a disruptive technology that delivers the future of additive manufacturing – creating high-value components with superior performance. Her passion has manifested into making wave(length)s in metal additive manufacturing since co-founding FormAlloy Technologies, Inc. in 2016. FormAlloy designs and manufactures award-winning metal deposition systems and solutions to a wide range of industries. Prior to FormAlloy, Melanie began her engineering career as an intern at Boeing and later spent 14 years as an engineer and program manager at Lockheed Martin. In 2020, she was named as one of SME's *20 Exceptional Women in Aerospace & Defense Manufacturing*. She holds a B.S. in aerospace engineering from the University of Illinois and an M.S. in systems architecture & engineering from the University of Southern California. In addition to her role at FormAlloy, Melanie currently serves on the America Makes executive committee, as the Vice President of Legislative Affairs for Navy League San Diego, and is the Women in 3D Printing Ambassador for San Diego.

Melanie credits her fascination with space since an early age, her enjoyment of math and science, and her affinity to the arts as the driving factor to becoming an engineer. Her desire to pursue a career that would hone her technical skills but allow her creative side to flourish led her to aerospace engineering. As a student, Melanie focused on space applications and designs for space tourism vehicles. Early in her professional career she focused on solving challenging technical problems and algorithm development for air defense systems. She quickly learned that as much as she loved the technical, side her favorite element of her career was working with people: learning, mentoring, and collaborating. She became involved with the local chapter of American Institute of Aeronautics and Astronautics (AIAA) to lobby for STEM and enjoyed Engineering Week projects where she first experienced the excitement of 3D printing. After relocating to different locations to take key roles in engineering, business development, and program management at Lockheed Martin, she made her home in San Diego. As a hobbyist, Melanie helped to build a simple polymer-based 3D printer in her own home and began attending "maker fairs." In 2016, she sought to fully harness the power of 3D-printing by working with metals and launched FormAlloy with her co-founder, Jeff Riemann. Since 2016, FormAlloy and their leadership team have won numerous awards and recognitions including the winner of Most Innovative Product for Defense and Cybersecurity, Project Entrepreneur, RAPID + TCT Innovation Audition Winner, Women in 3D Printing Innovation Award Finalists, 20 Exceptional Women in Aerospace & Defense Manufacturing, San Diego's 40 Under 40, and was named in the Top 5 startups in Additive Manufacturing in 2020 out of 143 that were analyzed.

Melanie resides in San Diego with her husband, three daughters, and dog named Cowboy. Outside of 3D printing and spending time with her family, she enjoys staying involved in the community, playing music, running, and yoga.

Chapter 9
From Substitution to Regeneration: The Tridimensional Interplay Between Cells and Biomaterials

Priscila Melo

The Overall Picture on Regeneration and the Perspective for Health in 3D: A Humble Opinion

The decades pass and the focus on the development of improved medical devices, which allow us to live longer with a better life-quality, has become unstoppable. Healthcare is a growing market, representing an open door for anyone willing to innovate and contribute to society. The gains are both economic and cultural as whilst we develop new ways to improve our health, we also make it more accessible to a larger slice of the population. In a utopian world, we would be able to remake our non-functional organs, which in the end could make humans immortal. Such ambitions are far from becoming a reality as our bodies are more complex than one might wish, but the development of technologies such as additive manufacturing (AM) and genetic programming have brought back the hope of organ substitution. This enthusiasm was firstly seen towards fields such as tissue engineering (TE), regenerative medicine (RM), and genetic therapy. AM and other novel biofabrication techniques have given an extra boost for their ability to create bespoke structures, with complex design, using new materials that are tailor-made for each processing technique, and application. The desire to go from substitution to regeneration is growing, and with it, the investment in these new therapies that are cell-based, or that use materials to stimulate cell function or even to create new tissues. The amount of research in TE and RM, associated with AM, has increased exponentially over the years, with target areas being orthopedics, oncology, and cardiology. Orthopedics is especially attractive as it encompasses problems that directly affect our daily lives, particularly with ageing, where bone regeneration is slower and pathologies such as osteoarthritis and osteoporosis begin to appear, and limit life in

P. Melo (✉)
Politecnico di Torino, Torino, Italy
e-mail: priscilamelo11@gmail.com

a significant manner. This attracts investment, translated in the growth of markets related to orthopedics, and materials used in it.

Biomaterials are essential for TE and RM therapies, being used to create platforms – scaffolds – that support and stimulate cells into proliferating and performing specific functions. The development of new materials is also stimulated by the availability of techniques able to process them into the desired designs. With AM, this is taken further as CT scans can be obtained from the patient directly, and the device can be fully personalized. Obviously, this encompasses a set of added costs which makes it harder to implement in a national health system. Such a level of personalization complicates the amount of devices that can be produced, which consequently affects the potential for scale-up. To find a middle ground, it is thought that investment should be focused on the materials that could be reshaped by surgeons directly in the operative room. Every approach is dependent on the needs of each pathology, which then influences the choice for the therapy and the tools used. Throughout the work developed on the PhD thesis represented in the following sections, the main aim was to find multiple uses and exploit the capacities of both biomaterial and manufacturing techniques, to broaden their window of applications within the orthopedic field.

Application of AM in Orthopedics and Tissue Engineering

The development of an adequate bone substitute, economical and bioresorbable, has been a major challenge for orthopedic, and reconstructive surgeons. Autografts have been the most common solution for bone defects, but with major drawbacks: (1) inconvenient for patients with degenerative bone diseases; (2) impractical for large defects as bone harvesting sites are limited [1, 2]. The current solutions for joint replacement or fracture fixation are mainly metal based, which leads to the problem of stress shielding, causing the bone to weaken and consequently to the failure of the implant [3]. Most metal-based implants are bioinert, and the cell survival on them is an issue, in addition to bacterial infection [4]. Such information turned the attention to other materials such as bioceramics and composites. New solutions were drafted based on these materials, as they are capable of mimicking bone and withstand the needed loads of the application site. These materials would be used as support and regeneration platforms and are the base of the new investigative fields of TE and RM [2, 5]. The access to AM techniques introduces a whole new range of possibilities, increasing the variety in terms of design, shape, and material choice for the scaffolds production [6]. For bone, glass-ceramics such as Apatite-Wollastonite (AW) have been used as scaffolds to study several pathologies as they combine bioactivity with mechanical strength [7–10]. Bioactivity is extremely important as it refers to the ability of the material to induce the precipitation of a carbonated apatite layer with a composition similar to the mineral phase of bone, hydroxyapatite (HA). This leads to an improved connection between implant and tissue, ensuring a successful fixation and integration. In terms of applications, AW has been used as a bone repair scaffold, implant, and filler [1, 9, 11–13]. AW can resist failure by fatigue under a constant loading of 65 MPa. This is higher than sintered HA which sustained the same loading for only 1 min. In the study developed by Kokubo et al. [14], it was estimated that AW can prevail for 10 years when submitted to those loading conditions, immersed in simulated body fluid (SBF) [15]. Bioactive materials are also being studied as reinforcement of polymer and metallic matrices of biocomposites to improve their fixation whilst enhancing osteoconductivity [16]. The potential of AW is yet underexplored which made it the focus of Dr. Melo's project. Her work exploited AW in different ways, resulting in two scientific publications [17, 18].

Defying Concepts and Exploiting Biomaterials for Load Bearing Applications

Apatite-Wollastonite in Bone Generation: The Next-Generation Bioceramics

AW is a glass-ceramic, meaning crystalline phases are formed within its glass matrix. In this case, apatite grains crystallize in a strong wollastonite matrix, via surface crystallization, which is the reason why it must be processed using sintering, a thermal treatment. Its processing is based on the melting of metal oxides, carbonates, and fluorides with the addition of modifiers to produce specific changes [19]. The method of producing AW scaffolds using binder jetting showed successful primary results, with mechanical properties mimicking those of cortical bone and full osseointegration, when implanted in rat calvaria [20].

Following the line of production for a medical device, Dr. Melo's project firstly investigates the raw material, AW, under the scope of load bearing applications. Focusing on the material development and thermal treatment, her interests were to (1) assess the impact of impurities, derived from the inherent steps of AW production, on its microstructure and process parameters, (2) determine how to avoid the impurities, and (3) recommend how to profit from such impurities. Once the effects associated with the contamination were outlined, the main question was whether this contaminant could be used as a dopant to facilitate the processing of scaffolds made with binder jetting, therefore reducing manufacturing costs. By studying it in depth, mainly the physics behind the phenomena associated with crystallization, Dr. Melo drew a doping strategy where such impurities were an advantage, when added in a controlled fashion. This study resulted in a scientific publication, which is summarized and discussed in the next section [18].

Alumina: From a Contaminant to a Dopant of Apatite-Wollastonite

Glass-ceramics are versatile materials since their properties are tunable at different levels. For example, the smallest of changes (below 2 wt% in glass composition) can affect parameters such as the bioactivity and degradation rate [21–23]. Within the medical field, such changes can dictate whether the material will be used for a temporary or permanent solution. Researchers such as Filho et al. [24, 25] demonstrated how the increase in crystallization can retard the development of the carbonated apatite layer. Introduction of oxides is often done on purpose to ease the processing of the material. In some cases, impurities result from the processing of the material itself, being alumina (Al_2O_3) amongst the most common ones. This oxide, which originates from the glass frit milling, acts as stabilizer. It has been used in many applications to control the microstructure during the sintering process of glasses, and other polycrystalline ceramics [26]. The AW processed in Dr. Melo's project was also subjected to ball milling, where the balls contained Al_2O_3. However, the small amounts detected were neglected up to the point where the changes in the scaffold processing were evident and connected to this impurity. Up to this point, the path to a final understanding was full of questions, answered by concepts that were not entirely understood. This made it a challenge to Dr. Melo, one worth pursuing as the interest was at both academic and industrial level. The first remark made by Dr. Melo stated that the control of material production should be taken to a higher level, starting with a decrease in milling time, and a change in the composition of the milling balls to zirconia. Once the problem was solved, Dr. Melo assessed the potential of the contamination as dopant, since knowing the potential of the materials, and exploring their properties, can open new possibilities towards the range of applications in which one can use them. Since AW is a glass-ceramic, its formation process is complex; therefore, it is key to establish the ideal processing parameters to obtain a repeatable manufacturing process [6, 27]. Dr. Melo's aim was to evaluate the effects of Al_2O_3 on the crystallization process of AW, and its implications on the sintering of bone implants. She focused on studying the effects of Al_2O_3 in the crystallization of AW, either as a surface contaminant originating from ball milling, or as a dopant, directly added in the initial formulation. Her main findings were that, when present at the surface of the particles, the alumina acted as a crystallization suppressor by occupying the nucleation sites of the AW precursor particles. When included in the initial batch composition, Al_2O_3 decelerated the material contraction whilst allowing for crystallization. This was reflected in the mechanical properties of the part which influenced the final application. Dr. Melo concluded that using alumina as a dopant was a useful tool, and a route that allowed for the processing of AW at lower temperature whilst maintaining its properties. Moreover, the presence of the dopant allows for higher control in the development of the microstructure, while maintaining crystallization and mechanical performance. This way it could be easily translated into an industrial environment, saving in processing costs, whilst keeping the main advantages of the material. The doping of glass-ceramics such as AW could be exploited, and the results of the study explored to further evaluate how the dopant could contribute for the control of crystal growth, resulting in its influence on the material mechanical properties and bioactivity in the presence of other oxides.

Binder Jetting of AW Scaffolds for Bone Regeneration: Is Crystallization a Blessing or a Curse?

The previous study on AW composition was a base investigation that would be used by Dr. Melo on the second part of her project: the manufacturing of bone scaffolds using binder jetting printing. She aimed to study the influence of crystallization on the physicochemical and biological properties of contaminated AW parts, produced with binder jetting followed by sintering. Processing of bone scaffolds implies a special attention to the geometry, mechanical properties, bioactivity, and biological potential. In this case, Dr. Melo aimed to understand the direct consequences of the material contamination in the mechanical properties of the resultant scaffolds, and their microstructure, as now the parts produced should present multilevel porosity. The mechanical properties were key as these scaffolds were meant for load bearing applications. The resultant microstructure had to be porous, with pore sizes ranging 200–300μm, to ensure cell migration within the structure [1]. Following her previous colleagues' work [3, 4], and applying her newly acquired knowledge, Dr. Melo created a powder blend consisting of 70% AW contaminated with 0.14 wt% Al_2O_3, and 30% Maltodextrin. Maltodextrin is a sugar that was used as a solid binder, an essential component in binder jetting of ceramics. This sugar dissolves when in contact with the liquid water-based binder, creating a glue that holds the AW particles together, allowing the formation of the initial part, the green body. This part is then taken to sinter, to attain the final structural integrity [1]. The binder jetting process adopted by Dr. Melo is described in Fig. 9.1. The raw powder is iteratively milled and sieved until the desired particle sizes are obtained. Maltodextrin is added to the AW powder and the materials mixed using a roller. The resultant blend is dried in a vacuum oven to remove residual humidity and then inserted in the printer feed bed. During the binder jetting process, a liquid binder is selectively deposited onto the powder bed in a layer-by-layer fashion. The resultant parts are left to dry overnight and then sintered in an oven, following an established protocol.

Dr. Melo intended on completing the blend catalogue developed by her group and to promote matter diffusion during thermal treatment, to improve the sintering of parts made with AW contaminated with 0.14 wt% Al_2O_3. She focused on using smaller particle size ranges to promote particle-particle contact and higher packing density. To evaluate the influence of the powder blend properties in the printing process, and the resulting green body, Dr. Melo assessed the sample's microstructure and mechanical properties alongside the bioactive character of the final scaffolds. Finally, she performed a biological assay to evaluate the osteoinductive and

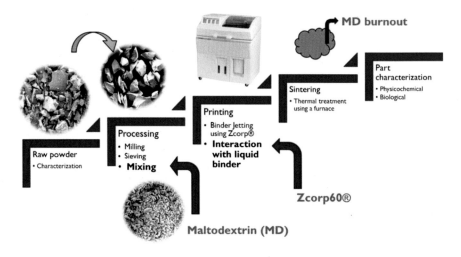

Fig. 9.1 Binder jetting of AW scaffolds, from powder blend formulation to scaffold manufacturing and finally thermal treatment. (Adapted from Melo [28])

osteoconductive abilities of the implants when in contact with human mesenchymal stromal cells (hTERT-MSCs). The results obtained in this study were paragoned to work developed by group colleagues who used an AW contaminated with 0.68 wt% Al_2O_3.

In terms of part processing, Dr. Melo observed that a successful manufacturing of the AW scaffolds depends on both powder blend properties and machine set-up/limitations. Despite the labelling of the blend as 'fair flowing', she concluded that a minimum amount of powder must be available to avoid layer dragging and excessive binder saturation. For the part characterization, the main observation relied on the relation between porosity and mechanical properties. She determined that, for glass-ceramics, the mechanical properties depend on both structural and crystalline properties of the material. For the same amount of porosity, and the same base composition, she observed a significant difference in Young's modulus, which was approximately 10 times higher for AW contaminated with 0.68 wt% Al_2O_3. She hypothesized that the difference in crystallinity, namely crystal size and orientation, proved to play a crucial part on the final properties of the parts. The degree of crystallinity was indirectly proportional to the amount of Al_2O_3, meaning AW contaminated with 0.14 wt% Al_2O_3 was more crystalline, proving once more that Al_2O_3 decreases the material's crystallization tendency. This increase in crystallinity influenced the thermal expansion coefficient (TEC) of the material, which expanded and contracted unevenly, resulting in the creation of stress points that later hindered the scaffold's mechanical performance. Allied to the layer defects, this led to a premature failure of the parts under loading, using 3-point bending test, which prevented the use of these scaffolds in load bearing devices. Dr. Melo suggested that the material processing must be finely controlled, especially the sintering step, and that an

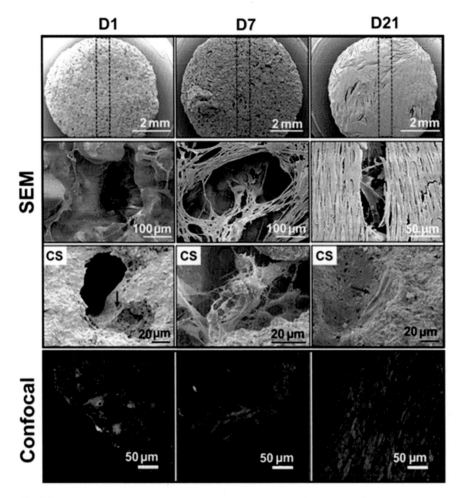

Fig. 9.2 Microstructure of AW scaffolds seeded with hTERT-MSCs at three time points D1, D7, and D21. SEM analysis demonstrated in the first three rows. Dashed rectangle indicates the cross-section area observed in micrographs tagged as CS. Cell morphology over 21 days, observed using confocal microscopy (GREEN: Vinculin for adhesion points; RED: Phalloidin for exoskeleton; BLUE: DAPI for cell nucleus). (Adapted from Melo [28])

annealing stage should be added to release internal stresses and further homogenize its distribution.

As for the bioactivity of the material, it remained unchanged, meaning the crystallization did not reduce the ability for ion exchange. The response of hTERT-MSCs was positive, with cells adhering and proliferating on the scaffolds produced, and differentiating into osteoblast within the first 7 days of culture (Fig. 9.2). This is evident from the SEM micrographs, where the cells multiplied to the point where they formed a tissue-like cape around the scaffold and the porous surface was no longer visible. At day 1 (D1) the presence of cells at the scaffold's surface was visible, with the cell attaching to the ceramic and the positions themselves inside the open pores. At this point, the cell's morphology was distinguishable and could be associated with a stem cell with its fibroblastic form. With time, the scaffold's surface was covered with cells as they proliferated and colonized. The cell's differentiation was also evident between day 1 and day 7 (D1, D7) as their shape transitioned from fibroblastic to triangular, typical of the transition from stem cell to preosteoblast (Fig. 9.2, confocal image D1 and D7). Despite the positive results at day 7, the reduced open porosity of the scaffold's structure hindered the cell migration, and even though cells reached the center of the scaffold, they were not able to survive for 21 days due to the lack of oxygen and nutrients. This was further worsened by the formation of a tissue-cape observed at day 21, which further blocked the entrance of nutrients.

Given these results, Dr. Melo concluded that if no further alterations were done regarding the binder jetting process and the thermal treatment, the scaffolds could be used in areas with lower loading, in small defects or as tissue regeneration platforms. For load bearing application, optimization of the printing process and powder blend is mandatory, alongside the addition of stabilizers and the inclusion of an annealing step.

With the development of different types of AW, Dr. Melo confirmed the influence of Al_2O_3 in part processing and its benefits when used as dopant. This is an added value for the industry since it allows for better regulation and quality inspection of the raw materials (glasses and glass-ceramic precursors) prior to commercialization. Also, this finding can aid the optimization of the manufacturing process so that it is possible to produce AW bone implants with different glass formulations resulting in a wide range of abilities, and possibly the use of other AM techniques such as selective laser sintering. As seen in literature [29, 30], crystallization benefits the mechanical properties of a part and therefore is desired for load bearing applications. However, Dr. Melo showed in her research that these previous findings are not directly translatable for glass-ceramics. The uncontrolled crystallization for glass-ceramics can indeed hinder the mechanical performance of a device. Despite the changes at the structural level, she showed that the devices were still osteoconductive, with an accelerated cell response towards differentiation and mineralization present from the first week of cell contact with the device.

Biocomposite Scaffolds and Stem Cells: The Next Step in Bone Regeneration

Dr. Melo has shown that AW is a versatile material but also noted that it had not yet been studied as reinforcement of polymer and metallic matrices of biocomposites, to enhance fixation and osteoconductivity [16, 31]. Since biodegradability is desirable, many polymers with mechanical properties within the range of bone have been explored, e.g. poly-L-lactic acid (PLLA). The reaction of the cells upon implantation is also under study, with the main goal being the translation from osteoconductive to osteoinductive [10, 32]. A material that is osteoconductive can support osteoblast growth and proliferation whilst being osteoinductive implies the differentiation of bone progenitor cells into osteoblasts. The addition of a bioactive material such as HA and AW is known to promote tissue growth adjacent to the implant and could possibly enhance the differentiation of human mesenchymal stem cells [33–35]. Up to date, studies on biocomposites using AM are limited, with most of them consisting of polymer matrices reinforced with calcium phosphate (CaP) particles [36]. Fiber-reinforced composites have proved to enhance mechanical properties whilst promoting osteoconduction [35, 37]. Multiple combinations can be used in terms of fiber configuration and length, but most studies have focused on long fibers, in a continuous or random configuration. For short fibers, the effect of this form on the bioactivity is unknown. However the influence of ion leaching from calcium-doped phosphate short glass fibers is reported on previous work done by Dr. Melo and it appears that it contributes to the biomineralization of hTERT-MSCs [38].

Fused Filament Fabrication (FFF): A Simple and Clean Method to Achieve Complex Results

To expand the use of AW as a bioactive filler, Dr. Melo developed a strategy that allowed the embedding of ceramic fillers within a polymer matrix which led to another scientific publication. In her work she chose PLLA, a well-known FDA-approved material, and added AW to the polymer matrix under the form of particles or short fibers. Dr. Melo's aim was to create an osteoinductive composite filament to manufacture scaffolds using FFF that would investigate the differentiation of hTERT-MSCs without the aid of osteoinductive factors [17]. Three types of porous woodpile samples were created: PLLA, PLLA with AW particles, and PLLA with AW fibers (Fig. 9.3). She found that the manufacturing of both filaments and

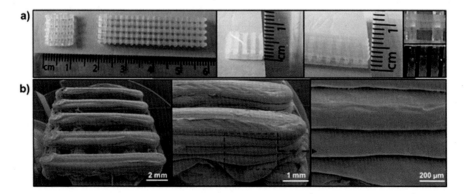

Fig. 9.3 PLLA scaffolds manufactured using FFF. (**a**) Photographs of scaffolds as produced. (**b**) Scaffold microstructure under SEM, with dashed squared highlighting the regularity and uniformity of the layers. (Adapted from Melo [28])

scaffolds was challenging, but once optimized, the produced structures were identical to the ones initially designed, including geometrical measurements and layer height (Fig. 9.3a). The SEM micrographs (Fig. 9.3b) prove that the layers were well deposited, and bond together, with some degree of melting, which is normal for this process and ensures a good adherence. Using these structures, Dr. Melo aimed to assess the influence of both AW fillers on their mechanical properties and degradation profile. Most importantly, based on her previous work, she predicted that the AW would render the scaffolds bioactive and therefore stimulate the differentiation of hTERT-MSC, followed by their mineralization. As an application, the scaffolds targeted bone repair in critical size defects, which are often overlooked and therefore in need of new solutions.

Spontaneous Stem Cell Differentiation and Mineralization

Cell differentiation was a goal, and according to the data reported on Fig. 9.4a, b, it was evidently achieved with these biocomposites. The increase in the concentration of alkaline phosphatase (ALP) was calculated in accordance with the number of cells (estimated previously) (Fig. 9.4a). The values indicate a significant increase in the concentration of ALP from day 1, for the PLLA samples and the biocomposite containing AW powder, indicating these samples are more osteoinductive. However, after 7 days, no differences were found between samples. This information is supported by the SEM analysis, where the change in shape from fibroblastic to cobblestone (Fig. 9.4b) illustrates the transition from stem cell to mature osteoblast.

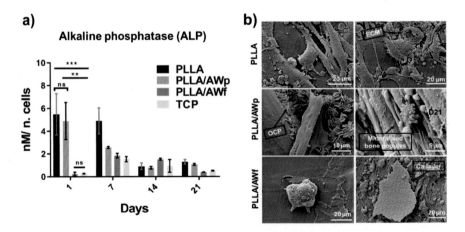

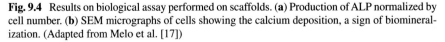

Fig. 9.4 Results on biological assay performed on scaffolds. (**a**) Production of ALP normalized by cell number. (**b**) SEM micrographs of cells showing the calcium deposition, a sign of biomineralization. (Adapted from Melo et al. [17])

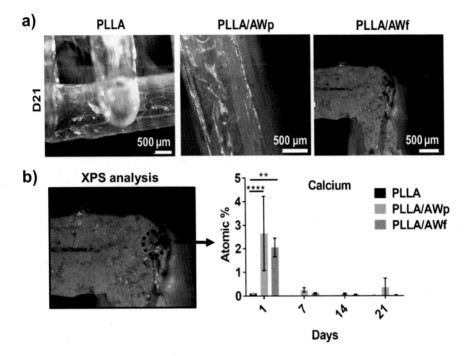

Fig. 9.5 Assessment of cell biomineralization. (**a**) Micrographs of alizarin red staining on scaffolds collected after 21 days after cell seeding. (**b**) Calcium deposition detected on top of the samples using x-ray photoelectron spectroscopy. (Adapted from Melo et al. [17])

The presence of calcium on top of the biocomposite samples indicated the material's ability to induce apatite deposition, which was detected with SEM (Fig. 9.4b) and alizarin red staining (Fig. 9.5a). The presence of red inferred a higher presence of calcium in the scaffolds seeded with cells, which was further confirmed with x-ray photoelectron spectroscopy analysis (XPS) (Fig. 9.5b). The analysis was performed in three areas of the scaffolds where an intense red coloration was seen as indicated by the dashed circle in Fig. 9.5b.

It was key to distinguish the calcium production arising from the material bioactivity from that of the cells. To achieve this, Dr Melo performed the alizarin red staining in samples immersed in simulated body fluid and when in contact with cells (Fig. 9.6). From the image it was clear the amount of calcium was higher in samples seeded with cells.

Observing calcium with and without cells is considered an asset since once implanted there would be two sources of calcium which would directly accelerate bone repair and regeneration. By combining these results with an ion leaching test, Dr. Melo demonstrated the importance of the calcium and magnesium ions in the cells' response. The amount and timing of the leaching influenced both cell proliferation and differentiation. She also proved that the filler's shape influenced the ion leaching profile, observing that the fiber's high aspect ratio allowed for higher

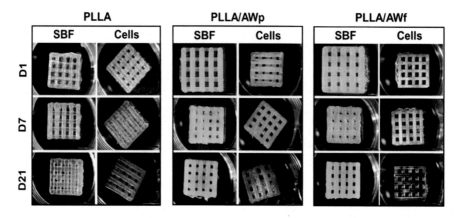

Fig. 9.6 Comparison between mineralization from immersion in simulated body fluid and from cells. (Adapted from Melo et al. [17])

exposure and longer leaching time, compared to the particulate biocomposite. This then reflected in the cells' response which was accelerated, but overall, both biocomposites promoted cell proliferation, differentiation, and mineralization. The mechanical properties were within the range of cancellous bone (50 MPa for bending strength), meaning both composites developed in Dr Melo's work could be used on critical size defects. Fiber-containing composites demonstrated higher ion leaching which rapidly stimulated cell proliferation and biomineralization, making them an interesting material to pursue.

Finding the Market Niche Within an Evolving Field

Healing of bone fractures and reconstruction of critical-size defects is a significant challenge since these do not heal spontaneously over the patient lifetime. Bone is a self-healing organ; however, this type of defect is highly dependent on two factors: (1) the surrounding tissue health, namely its vascularization to allow nutrient access; (2) the hTERT-MSCs presence in the bone marrow and periosteum, which are the osteoprogenitor cells. Secondary mechanisms influencing these cells include osteogenesis, osteoinduction, and osteoconduction. At the fracture site, osteoinductive factors are delivered by the vascular network. In defects where vascularisation is limited, an external stimulus must be applied to induce the cells differentiation [39]. In terms of osteoconduction within the human body, the osteoconductive platform is created via the formation of hematoma and cartilage callus, which then supports the osteoblast growth at the defect site [39]. Clinically, the most used treatments for critical size defects are bone grafts, distraction osteogenesis (DO), and induced membrane technique (Masquelet), but these all have their own limitations. From these three procedures, DO and Masquelet are the ones used in load bearing areas and suitable for long bone treatment due to the use of metal fixation systems. Bone grafts are used on their own, and in combination with DO and Masquelet, to promote faster bone regeneration at the defect site [39].

None of the implants developed in Dr. Melo's work reached the mechanical properties needed for load bearing areas, especially in long bone defects. For that reason, she focused on bone substitutes. Commercially available products exist mainly under the form of granules, cements or injectable pastes. The presence of antibiotics also limits their use, especially during pregnancy and nursing. There is a need for bone substitutes capable of supporting initial loading (minimum for cortical bone, ~50 MPa), without sourcing problems and size restrictions, and with an efficient manufacturing method. The production should allow for a versatile design able to be tailored to the defect site which is not as accurate when the device consists of a granular filler or a moldable cement and paste. The biocomposite scaffolds developed by Dr. Melo were intact after 8 weeks in degradation media, covering the period for initial bone repair (35 days) low loading areas such as maxilla. Other areas could be considered since the device mechanical performance is also dependent on the bone formation throughout the healing period. Bone remodeling rate is between 0.7 and 1μm/day [40], but these values are relative since the amount of remodeling depends on several factors such as age, site, surrounding tissue conditions and external stimulatory effects.

The scaffolds developed by Dr. Melo could also be of value for in vitro models, (e.g. disease models for both bone and joints), especially when bioceramic structures are combined with biocomposites. Previous work [41] using AW scaffolds and PLA hybrids attempted to mimic the bone tissue structure, where the porous bioceramic would act as the cortical fraction and the polymer as the trabeculae. In vivo testing showed that the hybrid structure improved bone growth when compared to AW. Also, it is suggested that the bioactivity of the AW influenced and enhanced cell migration and proliferation on the PLA fraction of the device. This opens another possibility for the use of the produced structures in the future, which are already valuable on their own.

Conclusion

The studies reported in this chapter show the work developed over the course of 3 years, during Dr. Melo's PhD project. She exploited different AM processes for making bone scaffolds, exploiting the potential of AW as a base material for bioceramic scaffolds and as filler in biocomposites. With the reported studies, Dr. Melo proved the importance of all steps in the manufacturing processing of medical devices. Despite the hype and excitement of AM and the curiosity to exploit its potential, she underlined that disregarding the material properties has harsh consequences and could lead to major problems in the part processing. It is important to recognize that the triad 'material/processing/properties' is key to obtain a successful outcome. It also provides more opportunities for innovation which, in the case of

Dr. Melo's work, was presented at all three levels: the material development, process optimization, and biological response. Dr. Melo evidenced the potential of using fillers with different forms, which resulted in the development of a new technique to blend the polymer and ceramics that promotes a homogeneous distribution of the filler within the matrix. She showed that osteoinductive polymer-ceramic biocomposites can be developed, using a small amount of inorganic phase (5 wt%), which up to date had not been reported. This method is now used by her former colleagues to create similar structures, using other polymers and ceramic fillers, processed by AM or traditional manufacturing.

Dr. Melo highlighted once more the importance of AM as a technique, and as a powerful tool in Industry 4.0. Being solvent-free techniques, both binder jetting and FFF have the potential to be scaled-up, following the directives of green manufacturing and supporting the guidelines for a more sustainable industry. In summary, the results obtained opened many doors. Several questions remain and must be answered, but most importantly, acquired concepts were demystified leading to what could be a new strategy for bone repair, completely tunable and with a high potential for commercialization.

References

1. Fujita, H., Iida, H., Ido, K., Matsuda, Y.: Porous apatite-wollastonite glass-ceramic as an intramedullary plug. J. Bone Jt. Surg. **82**, 614–618 (2000)
2. Wang, W., Yeung, K.W.K.: Bone grafts and biomaterials substitutes for bone defect repair: a review. Bioact. Mater. **2**, 224–247 (2017). https://doi.org/10.1016/j.bioactmat.2017.05.007
3. Shah, F.A., Thomsen, P., Palmquist, A.: Osseointegration and current interpretations of the bone-implant interface. Acta Biomater. **84**, 1–15 (2019). https://doi.org/10.1016/j.actbio.2018.11.018
4. Zhang, X.Z., Leary, M., Tang, H.P., et al.: Selective electron beam manufactured Ti-6Al-4V lattice structures for orthopedic implant applications: current status and outstanding challenges. Curr. Opin. Solid State Mater. Sci. **22**, 75–99 (2018). https://doi.org/10.1016/j.cossms.2018.05.002
5. Ratner, B.D., Hoffman, A.S., Schoen, F.J., Lemons, J.E.: Biomaterials Science: An Introduction to Materials in Medicine. Academic Press, Amsterdam (2004)
6. Butscher, A., Bohner, M., Hofmann, S., et al.: Structural and material approaches to bone tissue engineering in powder-based three-dimensional printing. Acta Biomater. **7**, 907–920 (2011). https://doi.org/10.1016/j.actbio.2010.09.039
7. Wopenka, B., Pasteris, J.D.: A mineralogical perspective on the apatite in bone. Mater. Sci. Eng. C. **25**, 131–143 (2005). https://doi.org/10.1016/j.msec.2005.01.008
8. Xiang, Z., Spector, M.: Biocompatibility of materials. In: Encyclopedia of Medical Devices and Instrumentation. John Wiley & Sons, Inc., Hoboken (2006)
9. Gomes, C.M., Zocca, A., Guenster, J.: Designing apatite-wollastonite (AW) porous scaffolds by powder-based 3D printing. (2015). https://doi.org/10.1201/b15961-30
10. Zwingenberger, S., Nich, C., Valladares, R.D., et al.: Recommendations and considerations for the use of biologics in orthopedic surgery. BioDrugs. **26**, 245–256 (2012). https://doi.org/10.1007/BF03261883

11. Kokubo, T., Kim, H.M., Kawashita, M.: Novel bioactive materials with different mechanical properties. Biomaterials. **24**, 2161–2175 (2003). https://doi.org/10.1016/S0142-9612(03)00044-9

12. Gomes, C.M., Zocca, A., Guenster, J., et al.: Designing apatite-wollastonite (AW) porous scaffolds by powder-based 3D printing. In: High Value Manufacturing: Advanced Research in Virtual and Rapid Prototyping, pp. 159–163. Taylor and Francis, Hoboken/London (2014). https://doi.org/10.1201/b15961-30

13. Ohtsuki, C., Kamitakahara, M., Miyazaki, T.: Bioactive ceramic-based materials with designed reactivity for bone tissue regeneration. J. R. Soc. Interface. **6**(Suppl 3), S349–S360 (2009). https://doi.org/10.1098/rsif.2008.0419.focus

14. Kokubo, T., Ito, S., Shigematsu, M., et al.: Mechanical properties of a new type of apatite-containing glass-ceramic for prosthetic application. J. Mater. Sci. **20**, 2001–2004 (1985). https://doi.org/10.1007/BF01112282

15. Duminis, T., Shahid, S., Hill, R.G.: Apatite glass-ceramics: a review. Front. Mater. **3**, 1–15 (2017). https://doi.org/10.3389/fmats.2016.00059

16. Navarro, M., Michiardi, A., Castan, O., Planell, J.A.: Biomaterials in orthopaedics. J. R. Soc. Interface. **5**, 1137–1158 (2008). https://doi.org/10.1098/rsif.2008.0151

17. Melo, P., Ferreira, A.-M., Waldron, K., et al.: Osteoinduction of 3D printed particulate and short-fibre reinforced composites produced using PLLA and apatite-wollastonite. Compos. Sci. Technol. **184**, 107834 (2019). https://doi.org/10.1016/j.compscitech.2019.107834

18. Melo, P., Kotlarz, M., Marshall, M., et al.: Effects of alumina on the thermal processing of apatite-wollastonite: changes in sintering, microstructure and crystallinity of compressed pellets. J. Eur. Ceram. Soc. **40**, 6107–6113 (2020). https://doi.org/10.1016/j.jeurceramsoc.2020.06.071

19. Kokubo, T.: Bioceramics and their clinical applications. Elsevier. ISBN 1845694 (2008)

20. Alharbi, N.H.: Indirect Three Dimensional Printing of Apatite-Wollastonite Structures for Biomedical Applications. Newcastle University, Newcastle upon Tyne (2016)

21. Abou Neel, E.A., Chrzanowski, W., Pickup, D.M., et al.: Structure and properties of strontium-doped phosphate-based glasses. J. R. Soc. Interface. **6**, 435–446 (2009). https://doi.org/10.1098/rsif.2008.0348

22. Miola, M., Brovarone, C.V., Maina, G., et al.: In vitro study of manganese-doped bioactive glasses for bone regeneration. Korean J. Couns. Psychother. **38**, 107–118 (2014). https://doi.org/10.1016/j.msec.2014.01.045

23. Murphy, S., Boyd, D., Moane, S., Bennett, M.: The effect of composition on ion release from Ca-Sr-Na-Zn-Si glass bone grafts. J. Mater. Sci. Mater. Med. **20**, 2207–2214 (2009). https://doi.org/10.1007/s10856-009-3789-y

24. Filho, O.P., La Torre, G.P., Hench, L.L.: Effect of crystallization on apatite-layer formation of bioactive glass 45S5. J. Biomed. Mater. Res. **30**, 509–514 (1996). https://doi.org/10.1002/(SICI)1097-4636(199604)30:4<509::AID-JBM9>3.0.CO;2-T

25. Massera, J., Mayran, M., Rocherullé, J., Hupa, L.: Crystallization behavior of phosphate glasses and its impact on the glasses' bioactivity. J. Mater. Sci. **50**, 3091–3102 (2015). https://doi.org/10.1007/s10853-015-8869-4

26. Baino, F., Marshall, M., Kirk, N., Vitale-brovarone, C.: Design, selection and characterization of novel glasses and glass-ceramics for use in prosthetic applications. Ceram. Int. **42**, 1482–1491 (2016). https://doi.org/10.1016/j.ceramint.2015.09.094

27. Butscher, A., Bohner, M., Doebelin, N., et al.: Moisture based three-dimensional printing of calcium phosphate structures for scaffold engineering. Acta Biomater. **9**, 5369–5378 (2013). https://doi.org/10.1016/j.actbio.2012.10.009

28. Melo, P.: Additive Manufacturing of Bioceramic and Biocomposite Devices for Bone Repair. Newcastle University, Newcastle upon Tyne (2019)

29. Freiman, S.W., Hench, L.L.: Effect of crystallization on the mechanical properties of Li2O-SiO2 glass-ceramics. J. Am. Ceram. Soc. **55**, 86–90 (1972). https://doi.org/10.1111/j.1151-2916.1972.tb11216.x

30. El-Kheshen, A.A., Khaliafa, F.A., Saad, E.A., Elwan, R.L.: Effect of Al2O3 addition on bioactivity, thermal and mechanical properties of some bioactive glasses. Ceram. Int. **34**, 1667–1673 (2008). https://doi.org/10.1016/j.ceramint.2007.05.016

31. Juhasz, J.A., Best, S.M., Brooks, R., et al.: Mechanical properties of glass-ceramic A–W--polyethylene composites: effect of filler content and particle size. Biomaterials. **25**, 949–955 (2004). https://doi.org/10.1016/J.BIOMATERIALS.2003.07.005

32. Winkler, T., Sass, F.A., Duda, G.N., Schmidt-Bleek, K.: A review of biomaterials in bone defect healing, remaining shortcomings and future opportunities for bone tissue engineering. Bone Jt. Res. **7**, 232–243 (2018). https://doi.org/10.1302/2046-3758.73.BJR-2017-0270.R1

33. Boccaccini, A., Maquet, V.: Bioresorbable and bioactive polymer/bioglass composites with tailored pore structure for tissue engineering applications. Compos. Sci. Technol. **63**(16), 2417–2429 (2003)

34. Wang, M.: Developing bioactive composite materials for tissue replacement. Biomaterials. **24**, 2133–2151 (2003). https://doi.org/10.1016/S0142-9612(03)00037-1

35. Ahmed, I., Jones, I.A., Parsons, A.J., et al.: Composites for bone repair: phosphate glass fibre reinforced PLA with varying fibre architecture. J. Mater. Sci. Mater. Med. **22**, 1825–1834 (2011). https://doi.org/10.1007/s10856-011-4361-0

36. Yasa, E., Ersoy, K.: Additive manufacturing of polymer matrix composites. Composites, Aircraft Technology, Melih Cemal Kuşhan, IntechOpen (2018). https://doi.org/10.5772/intechopen.75628. Available from: https://www.intechopen.com/books/aircraft-technology/additive-manufacturing-of-polymer-matrixcomposites

37. Felfel, R.M., Ahmed, I., Parsons, A.J., et al.: Cytocompatibility, degradation, mechanical property retention and ion release profiles for phosphate glass fibre reinforced composite rods. Mater. Sci. Eng. C. **33**, 1914–1924 (2013). https://doi.org/10.1016/j.msec.2012.12.089

38. Melo, P., Tarrant, E., Swift, T., et al.: Short phosphate glass fiber – PLLA composite to promote bone mineralization. Mater. Sci. Eng. C. **104**, 109929 (2019). https://doi.org/10.1016/j.msec.2019.109929

39. Roddy, E., Debaun, M.R., Daoud, A., et al.: Treatment of critical-sized bone defects: clinical and tissue engineering perspectives. Eur. J. Orthop. Surg. Traumatol. **28**, 351–362 (2018). https://doi.org/10.1007/s00590-017-2063-0

40. Burr, D.B., Gallant, M.A.: Bone remodelling in osteoarthritis. Nat. Rev. Rheumatol. **8**, 665 (2012)

41. Rodrigues, N.: Materials Processing and Physical Characterisation of a Hybrid Composite Structure for Bone Replacement Applications. Newcastle University, Newcastle upon Tyne (2018)

Dr. Priscila Melo is a biomedical engineer, specialised in additive manufacturing (AM) of biomaterials for regenerative medicine. Her academic path started with a bachelor's degree in biomedical engineering, followed by a master's degree where she specialised in biomaterials and medical devices. Her work in regenerative medicine started then, with the study of PLLA films, polarized using a corona discharge, for the guidance of osteoblasts during the bone healing process. Being passionate about research, this led to the next step in her career, a PhD in AM, one of the first degrees in the area created by the Centre for Doctoral Training from the EPSRC (United Kingdom). This made her one of the first experts to be formally graduated in the area, applied to in medical field. During the PhD, her research mainly focused on the development of polymer-ceramic composites and glass-ceramics, used for the production of devices able to support bone growth and induce differentiation of bone marrow–derived human stem cells (hMSCs) in basal media. These devices were processed using different AM techniques and all of them were tested biologically, to certify their osteoinductive and osteoconductive abilities. Wishing to expand her knowledge, she accepted to head and execute the main activities of a project aiming the development of polymer-ceramic craniofacial fracture fixation devices, using compression moulding (traditional manufacturing). Being an industry-sponsored degree, she also had the opportunity to work within the company that was part of the project and help them develop and optimise the materials supplied for the PhD thesis. Within this context she contributed to the solution of a synthesis problem and, by further studying the physics of the produced glass-ceramic, was able to understand how it behaved and to optimise its processing into bone scaffolds via binder jetting (AM). At the end of the PhD, her objective to become a versatile and multi skilled professional was achieved. She had gained experience in all steps of making a medical device using AM, from material development, process optimisation, and characterisation, and had become independent in biological testing, despite it not being part of her academic formation.

Her work within the field of AM was considered new and promising and was presented at prestigious conferences in the field. This expertise and ability led to her invitation as chair in several events, and also session organiser, despite being in an early stage in her career. Dr. Priscila Melo has also been part of the judge committee for the MIT project competition 'Innovators under 35', amongst other experienced and renowned professionals in area, being one of the youngest guests to ever participate.

Being a proactive member of the scientific community, she was also involved in seminars aiming to promote AM within the health sector to both academic and general public, being part of the Women in 3D printing community.

After graduating, she followed the research path initiating a new learning cycle at the IRIS group, at Politecnico di Torino, working with natural polymers and composites, processed using AM and electrospinning.

Chapter 10
Ceramic Additive for Aerospace

Lisa Rueschhoff

Introduction: Ceramics for Aerospace and the Need for Additive Manufacturing

The cruise altitudes and speed of flight will be higher than ever before for next-generation aerospace applications. While this will enable superior efficiency and reach, especially for military aerospace vehicles, it comes at the cost of much harsher environments experienced by the materials used to produce these components. Not to mention modern modeling has enabled the discovery of more efficient and complex geometrical part designs that are difficult to make with conventional methods.

Ceramic materials are of interest for these applications since they can withstand higher temperatures and harsher environments than many traditional metal or polymer aerospace components. Beyond increased temperature capability, they also offer increased erosion resistance, higher stiffness, lower density, and, in some cases, multi-functional properties. However, the brittle nature of monolithic ceramic components often limits their use in high stress environments. As such, ceramic materials either are limited to non-critical structural components or must be reinforced to enhance fracture toughness. The use of fiber reinforcements to form ceramic matrix composites (CMCs) is one way to improve the mechanical performance and provide more graceful failure and enhanced load bearing capabilities.

One prime example of the promise of ceramics for aerospace is the adaptation of CMCs in commercial jet engines. Silicon carbide (SiC) fiber-reinforced SiC matrix (SiC/SiC) CMCs have begun replacing nickel-based super alloys in both stationary

The original version of this chapter was revised. The correction to this chapter is available at https://doi.org/10.1007/978-3-030-70736-1_13

L. Rueschhoff (✉)
Air Force Research Laboratory, Dayton, OH, USA
e-mail: lisa.rueschhoff.1@us.af.mil

© Springer Nature Switzerland AG 2021, Corrected Publication 2021
S. M. DelVecchio (ed.), *Women in 3D Printing*, Women in Engineering and Science, https://doi.org/10.1007/978-3-030-70736-1_10

and rotating components in the hot section of commercial engines [1]. This replacement comes with a ~250 °C increase in operating temperature and a one-third reduction in weight [1]. While SiC/SiC CMCs are finding success in components of relatively simple geometries, they are difficult and costly to shape into complex shapes. Conventional CMC processing methods used to infiltrate the ceramic fiber preforms, such as polymer or chemical vapor infiltration, are time consuming, costly, and limited to relatively simple geometries. Ceramic monoliths are somewhat easier to process but are still limited in terms of near-net complex shaping techniques.

Additive manufacturing (AM) of ceramics offers a more agile manufacturing method to create complex-shaped components with the potential for spatial microstructural control [2]. AM has been gaining momentum across all materials in recent years, with most advancements and industrial implementation being in the field of polymers and metals. Due to the complexities that come with forming dense ceramic materials, the field of AM is still in initial stages of adaptation. A variety of AM routes exist for forming ceramic materials, each with their own advantages and limitations. This chapter will briefly introduce the background and context of ceramic AM techniques. More detailed information on this topic can be found in many excellent recent review articles [2–5]. The focus of the chapter will be to highlight my recent advances in AM of ceramics and CMCs using the technique of direct ink writing. This will give just a few examples of many current research level efforts that are laying the critical foundation in this rapidly growing field.

Additive Manufacturing Methods for Ceramics

The use of additive manufacturing for fabrication of ceramics has grown rapidly in the past decade [4–7]. The techniques used for ceramic AM generally build on those developed for polymer and/or metallic materials. In fact, most equipment used to additively manufacture ceramic parts is marketed for other materials (i.e., polymers or metals) with few examples of equipment designed and sold with ceramics in mind. A main differentiator in the adaptation of these AM methods to ceramics is the necessity for sintering or densification of the ceramic post-processing. Nearly all ceramic AM techniques result in a "green" ceramic part that must be fully densified to achieve robust mechanical performance.

In this section, each ceramic AM method will be briefly described for context. The methods can be separated into categories depending on the feedstock form (see flow diagram shown in Fig. 10.1). While each technique has been explored for the use of complex-shaped ceramic components, the work is still primarily on the research scale with some examples of commercial production. This is expected to change in the next ~5 years as advancements continue to push forward in all areas.

Powder-based AM processes involve a ceramic powder bed that is built up layer-by-layer and held in place through either liquid binder droplets (in the case of binder jet printing) or partial to full melting/sintering using a laser source (selective laser melting or sintering). These technologies became popular and have advanced through

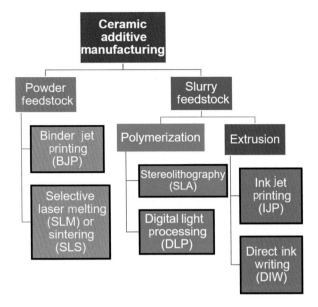

Fig. 10.1 Flow chart of ceramic-based additive manufacturing technologies

powder metallurgy AM and have been adapted to ceramic powders. In both techniques, a thin layer of ceramic powder is spread across the build platform using either a blade or roller. Flow of the ceramic powder is an important feat to overcome in order to achieve dense starting layers. More information on the binder jetting technique can be found in the glossary. Laser-based methods, such as selective laser melting or sintering (SLM or SLS, respectively), are advantageous as they can produce either fully or partially sintered ceramic components. During SLS, the laser heats the material up to a specific point below the melting temperature such that the particles are only partially melted to fuse together to hold their shape [4]. Post-processing steps such as reactive melt infiltration [8, 9], chemical vapor infiltration [10], or slurry infiltration and sintering are needed to fully densify the part. Lasers with much higher energy density are used for SLM in order to fully melt the ceramic powders, and as a result it is the only ceramic AM process to result in dense ceramic components. The method has been used successfully for ceramic coatings [11], but issues with thermal shock make it difficult to build up complex parts. While the powder feedstock is relatively low in cost depending on the composition, the high cost of equipment can be a barrier for some academic research labs or small businesses.

Slurry-based AM methods build off of years of knowledge developing solvent-based ceramic slurries for conventional processing techniques like slip or tape casting. Polymerization routes rely on polymer binder curing using a UV-light source for shape retention of individual layers which are then built up layer-by-layer to form a complex shape. Stereolithography (SLA) and digital light processing (DLP) are the most common techniques that have been explored. To cure each layer, the UV-light source is either scanned point-by-point (in the case of SLA) or projected all at once through a patterned mask (DLP) [12–17]. Slurries generally are solvent based and contain fine ceramic powders (~nm/μm) dispersed in a UV-curable polymer binder phase [4]. Exploratory work shows that using preceramic polymers rather than ceramic powders can enhance feature resolution [18, 19]. There has been great success with certain ceramic phases (e.g., silica, alumina) that have amenable properties such as low index of refraction as to not scatter UV light [20, 21]. Success and interest in this technique has led to the production of mass-produced printers marketed specifically for ceramics (e.g., Lithoz, Admatec, and Tethon). High resolution control and surface quality are advantages of this process, but come at the cost of limited ceramic powder compatibility and build platform size.

In slurry feedstock extrusion-based methods (Fig. 10.1), a ceramic slurry is deposited in either droplet or continuous filament form in a layer-by-layer sequence [22]. Droplet-based methods include direct inkjet printing, where droplets are deposited into a desired pattern in a layer-by-layer sequence. Ceramic inks for this process contain a high amount of solvent and a minimal ceramic powder content (~5 vol.%). This low solids content provides the viscosity needed for printing, but limits the shape retention that can be achieved which is necessary for the build-up of complex shapes. Since the process is not continuous, total forming time is limited by the time to print each droplet. Conversely, filament-based approaches such as fused deposition, freeze-form extrusion fabrication, free-form fabrication, and direct ink writing utilize a continuous extrusion of highly concentrated ceramic inks.

Of particular interest is direct ink writing (DIW), where ceramic-based inks are extruded through a nozzle in a layer-by-layer sequence (Fig. 10.2). The technique has been used to print a variety of ceramic structures and has gained popularity due to the low cost of equipment and ability to use a variety of ink feedstocks including solvent-, aqueous-, or preceramic polymer–based [12, 13, 23–26]. The resolution or feature size of the process depends on the size of nozzle used and has varied from 0.1 to 1.5 mm [6, 27]. The surface roughness of the printed parts depends on the preset layer height (the amount the build platform lowers between each deposited layer) and the rheology of the ink. The surface quality and resolution of final parts

Fig. 10.2 Schematic of DIW process

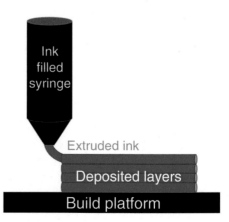

depends on the print parameters (e.g., nozzle size, layer height, ink rheology) but are generally lower than those produced via SLA/DLP.

First iterations of the technique utilized aqueous- or solvent-based inks that were highly loaded (>45 vol.%) with ceramic powders, yielding near-net shaped parts after pressureless sintering [12, 13, 22, 27]. Recently, preceramic polymer (PCP)–based inks have been developed and successfully implemented in DIW [14–17]. The commercially available PCPs that are used act as a flowable medium for ceramic particle or other filler material. Rather than burning away like other polymer binders and dispersants used in aqueous- or solvent-based ceramic inks, the PCPs convert to ceramic material upon high-temperature pyrolysis (>800 °C) [28]. There are a variety of compositions available in preceramic polymer systems (e.g., SiC, SiOC, Si_3N_4 etc.), as well as research efforts in the synthesis of custom chemistries [28, 29].

Regardless of formulation, the rheology of flow behavior is a key parameter in the printability of the inks. A simplified schematic illustrating different types of rheological behavior is shown in Fig. 10.3. Traditional Newtonian fluids exhibit a constant viscosity (slope of the shear stress vs. shear rate) over all applied shear rates. Pseudoplastic or shear-thinning materials are more commonly observed in complex fluids and display a decrease in viscosity with increased applied shear rate. Most desirable for DIW is yield-pseudoplastic behavior which requires overcoming a yield stress (τ_y) before pseudoplastic flow [12, 13]. Pseudoplastic behavior is advantageous for DIW as the applied load during printing assists in decreasing the viscosity of the ink, making it easier to extrude. The yield stress of the ink enables the material to retain the extruded filament shape once the stress is removed, allowing the deposition of multiple stacked layers [13]. The yield stress comes from inherent particle flocculation that occurs in particle-loaded solvent inks due to steric and/or electrostatic attraction. A loose network forms, inhibiting flow without applied stress. The stronger the attraction, the higher the yield stress and more difficult it is to overcome during the printing process. For weak particle interactions,

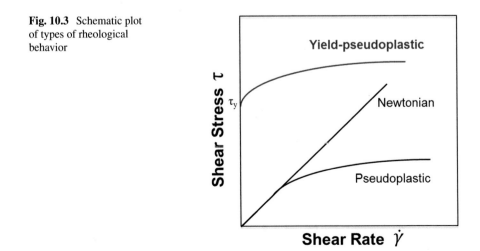

Fig. 10.3 Schematic plot of types of rheological behavior

the yield stress is not high enough to support the weight of additional printed layers causing a flattening of bottom layers and loss in shape retention. An ideal ink has a yield stress that can easily be overcome through the printing process which can depend on the nozzle size and pressure or torque capabilities of the printer. Within a certain developed ink, rheological modifications are achieved through varying ink constituents, including but not limited to: ceramic powder solids loading, polymer dispersant/binder concentration or molecular weight, and solvent selection.

Personal Advancements in Direct Ink Writing

Ceramic Monoliths

My first efforts in utilizing DIW to create complex-shaped ceramic components started as a side project during my PhD. We purchased a low-cost syringe style 3D printer that was marketed toward printing chocolate and adapted it to extrude ceramic suspensions. I executed this work as part of a team of researchers at Purdue University including: Prof. Rodney Trice, Prof. Jeffrey Youngblood, Dr. William Costakis, et al. [13]. Alumina was targeted for first experiments as a model material due to the ease of densification and feedstock powder low cost and availability. Moreover, highly loaded alumina aqueous suspensions had been previously developed in our group for a modified injection molding processes that could easily be adapted to the DIW process [30, 31]. These suspensions were composed of alumina ceramic powder (>50 vol.%), water-soluble polymer dispersant, and polymer binder. An ammonium polyacrylate solution (PAA-Na, tradename Darvan 821A) was identified in previous studies as an effective dispersant for particle stabilization [31]. This improved stability affords efficient particle packing and, therefore, ability to increase solids loading while maintaining a low viscosity. Polyvinynlpyrolidone (PVP) with a 55,000 g/mol molecular weight was selected as a binder to increase both rheological yield strength and green body machinability [31, 32].

In this study, we investigated the adaptation of these alumina suspensions as inks for DIW with the underlying design principle of maximizing solids loading while maintaining requisite rheology and printability. Five total suspensions were developed as DIW inks with alumina loading ranging from 51 to 58 vol.%, with nominally constant dispersant and binder content (approximately 4 vol.% and 5 vol.%, respectively). The effect of alumina solids loading on rheology and printability, as well as the resulting physical (i.e., density and microstructure) and mechanical properties of the sintered parts, was evaluated.

The processing steps to achieve final densified parts are outlined in Fig. 10.4. The inks were all mixed to achieve desired composition and were loaded into a standard 10 mL syringe with a luer lock tip (ID = 1.25 mm) for printing. After printing, parts were left at room temperature overnight to adequately remove water. Binder burnout was conducted at 700 °C in air in order to remove the polymer phases that

Fig. 10.4 Processing steps for the direct ink writing of aqueous-based ceramic slurries

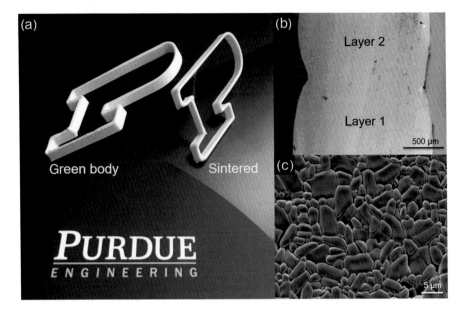

Fig. 10.5 Montage of DIW alumina samples. (**a**) Green body and sintered hollow Purdue "P" shape made from DIW. (**b**) Cross-section of sintered component with (**c**) higher magnitude SEM image of microstructure. (Adapted from Ref. [13])

degrade at elevated temperatures [31]. Finally, parts were sintered, without the use of external pressure, at 1600 °C in air to remove residual porosity and achieve a densified, robust ceramic part.

While all five inks exhibited requisite rheology for printing, the printability of the 53–56 vol.% inks were most ideal. The lowest solids loading ink, at 51 vol.%, had the lowest yield stress and viscosity. As a result, the part experienced slumping after printing as the originally extruded layers were unable to support the weight of subsequent layers. At the highest solids loading of 58 vol.%, the high viscosity and low water content led to nozzle clogging and clumping of the ink during printing. This resulted in non-uniformity of the printing layers and part defects. The ink containing 55 vol.% alumina was deemed optimal for the printing process as it had the most uniformity in printed layers. Parts printed with this optimized ink are shown in Fig. 10.5. An as-printed green body and sintered shape are shown in Fig. 10.5a and exhibit excellent part fidelity as evidenced through fine lines and sharp edges of the parts. Figure 10.5b shows a cross-section of a sintered part to illustrate the adhesion between subsequent printed layers, with no voids or cracks appearing at the interface. Finally, the microstructure shown in Fig. 10.5c is as expected for alumina, with an average grain size of nominally 3.5 μm. All samples were sintered to a minimum of 98% theoretical density for the material, with the remaining ~2% porosity attributed to air introduced during ink mixing and filling of the syringe. The average flexural strength, as measured using 3-point bend testing, ranged from 134 to 157 MPa. These values are similar or slightly lower than others reported in literature for pressureless sintered alumina, likely due to the homogenously distributed pores seen throughout the cross-section that act as crack initiation sites.

This study on alumina DIW helped lay the groundwork for low-cost, agile manufacturing of ceramics through DIW of aqueous-loaded suspensions. Follow-on work from our team included application to the boron carbide system (for interest as an armor material) and some work on silicon nitride (unpublished) [12]. Silicon nitride is one of the few monoliths that have been explored for use in aerospace applications due to its temperature resistance paired with enhanced fracture toughness from an interlocking microstructure [33]. However, for more critical structural aerospace components, higher toughness and more graceful failures than monolithic phases can provide are generally needed.

Ceramic Matrix Composites

The addition of fiber reinforcements to monolithic ceramic components, in the form of continuous or chopped fibers, is used to combat the intrinsic brittle nature of ceramics. Continuous fiber reinforcements offer the greatest increase in fracture toughness but are more difficult to adapt to existing AM processing techniques. A few exceptions exist in literature where continuous SiC or SiO_2 fibers were incorporated in extrusion-based AM surrounded by a ceramic-based matrix [24, 34]. Research continues in this area but is still in its infancy. Discontinuous reinforcements, such as chopped fibers or whisker phases, provide an easier route for incorporation into existing feedstocks for slurry- or powder-based processing. While they do not impart as much mechanical performance as continuous fibers, research has shown an increase over monolithic ceramic materials [35–40]. Primarily traditional ceramic processing methods have been used to produce chopped fiber composites, such as powder pressing of dry powder mixtures followed by sintering for densification. In general, these techniques produce parts with low geometric complexity and lack of fiber alignment for enhanced mechanical performance.

Utilizing DIW for chopped fiber composites can allow for complex-shaped components with spatially controlled microstructure (i.e., localized fiber alignment). The shear forces in the nozzle during printing can be used to align high aspect ratio reinforcement phases to increase mechanical performance. Pioneering work from Compton and Lewis on DIW of SiC whisker and carbon fiber (C_f) filled epoxy composites has proven the feasibility of alignment of these reinforcing phases during the forming process [41]. Using this process, the authors were able to create aligned microstructures with up to a 9x increase in Young's Modulus compared to base epoxy samples [41]. The same methodology has since been applied by Franchin et al. to a CMC system [16]. In their work, they showed preliminary success of printing C_f/SiOC composites with a chopped fiber filled preceramic polymer ink [16].

Along with a team of researchers from AFRL (Dr. Zlatomir Apostolov), University of Tennessee-Knoxville (J. William Kemp and Prof. Brett Compton), we have adapted this approach to print CMCs with an ultra-high temperature ceramic (UHTC) matrix phase [23]. UHTCs are a class of boride, carbide, and nitride ceramics with extreme temperature capabilities (melting temperatures >3000 °C) and environmental stability. As such, they have been explored as high-temperature coatings or even as entire leading edge components for high-speed flight applications, where the sharp leading edge can reach in excess of 2000 °C [42]. However, practical use as a leading edge is still limited by poor fracture toughness and resulting limited thermal shock resistance [43]. In fact, a report on monolithic UHTCs tested in simulated atmospheric re-entry conditions often found failure from catastrophic brittle fracture during testing due to thermal shock [44, 45].

Reinforcing UHTCs with a secondary fiber phase have been explored as a means to increase fracture toughness and alleviate catastrophic brittle failure [36, 37, 43, 46–52]. These UHTCMCs offer the temperature and oxidation resistance of UHTCs, along with the strength, fracture toughness, and strain to failure of a composite. The use of a UHTC matrix allows for higher temperature capability and improved ablation performance over traditional aerospace CMCs (e.g., C/C or SiC/SiC) [53, 54]. A unique capability of the DIW process is the ability to spatially control composition and alignment of fibers where it is specifically needed, such as in high stress areas during use or around a geometric stress concentrator. The area of processing of UHTCMCs is rapidly growing, but no work has been done to produce them using additive manufacturing.

For our work, we developed novel DIW inks containing a SiC-yielding preceramic polymer (PCP), UHTC zirconium diboride (ZrB_2) powder, and chopped SiC fiber (SiC_f) [23]. The PCP used for this study is a commercially available polycarbosilane (SMP-10, Starfire Systems) with a ceramic yield of ~80 wt.% after high-temperature pyrolysis [55]. The phase of ZrB_2 was chosen due to the lower cost and density compared to other UHTCs. The added SiC_f reinforcement was cut into roughly 1 cm lengths before adding to the mixture and was milled during ink mixing to a final average fiber length of ~0.35 mm. All ink components were thoroughly mixed using a planetary type mixer before the printing process.

Three inks were developed with requisite rheology for printing, all with nominally 43 vol.% solids loading (ZrB_2 + SiC_f) and varying levels of SiC_f (0, 5.75 and 10 vol.%). All three inks were used to print simple hollow boxes and more complex honeycomb-type lattice structures (Fig. 10.6a) using a displacement controlled DIW printer (Hyrel 30M) with a 0.84 μm nozzle diameter. The processing steps are similar to those reported in the previous section for aqueous slurries (Fig. 10.4), but

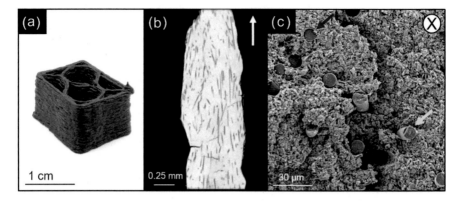

Fig. 10.6 (**a**) As-printed component made from 5.75 vol.% SiC$_f$ ink, (**b**) X-Ray CT cross-section (arrow denoting print direction), and (**c**) SEM image of fractured surface. The arrow going into the page at the top right denotes the print direction while the red and yellow arrows point out examples of a fiber break and fiber pull-out, respectively

with polymer curing and pyrolysis steps for densification instead of air dry, binder burnout, and pressureless sintering. The thermal curing at 230 °C after printing acts to cross-link and solidify the PCP, followed by pyrolysis at 1200 °C to convert the polymer to ceramic. X-ray computed tomography (XCT) was used to visualize fiber alignment in the printed part, with an example from the inks with 5.75 vol.% fiber (Fig. 10.6b). Strong alignment is observed in the print direction with an average fiber angle of 89.2 ± 20.2° (where the print direction = 90°), as manually measured from 5 total XCT images of nearly 1000 individual fibers. A slightly lower degree of alignment and a greater standard deviation (83.0 ± 32.7°) was observed for the parts printed with a higher content of SiC$_f$ in the ink (10 vol.%). This may be attributed to the increase in fiber-fiber interactions, as observed from others in polymer chopped fiber composite systems [56, 57].

The mechanical behavior of UHTCMC samples made from all three inks was tested using a 3-point bend test, with a representative fracture surface shown in Fig. 10.6c. While evidence of fiber pullout (pointed out with a yellow arrow) is observed in the fracture surface, there are also examples of fibers that have fractured in the crack plane (red arrow). The fiber pullout is a known toughening mechanism for ceramic composites and is an indicator of a weak fiber-matrix interface. The mechanical strength determined from these tests (<60 MPa bending strength) is much lower than expected but is attributed to porosity observed in the samples. This porosity is detrimental to ceramic strength and is due to the preceramic polymer out-gassing during the cross-linking reaction, along with cracking that can occur from shrinkage during conversion to ceramic. Porosity is commonly observed in polymer-derived ceramics and is often mitigated through multiple infiltration cycles (e.g., the traditional CMC processing technique of polymer infiltration and pyrolysis, a.k.a. PIP) [55, 58]. Our preliminary experiments infiltrating PCP into the printed parts resulted in up to 6% weight increase, demonstrating promise that will be explored in future efforts.

This work is the first report on fabricating UHTCMCs using additive manufacturing and provides the foundation to further expand AM for this class of materials for high-temperature, high-speed flight applications. Current and future avenues to explore include developing an enhanced understanding and optimization of the UHTCMC direct ink writing process, especially as it relates to the structure-property relationships. This includes exploring the processing space of printing parameters and the effect on fiber alignment and subsequent mechanical properties. Developing an understanding of this processing space will enable more tailored development of components based on desired mechanical performance.

Summary and Future Outlook

The field of additive manufacturing of ceramics is rapidly growing with new technology emerging frequently. My research in the last 5 years has been to utilize the technique of direct ink writing to produce robust, complex-shaped ceramics and CMC components. These materials show promise across a variety of high-temperature aerospace applications that are in the need for more agile manufacturing methods. This work helped, in part, to lay some of the groundwork for adapting this technique across a plethora of ceramic materials. It is expected that future avenues to explore for ceramic DIW include in-line mixing for functional graded structures and expanded utilization of shear alignment of reinforcement phases such as whiskers, chopped fibers, and platelets [2, 59]. As research continues to emerge showing the promise of ceramic additive manufacturing, widespread adaptation into industrial processes is expected to follow. Continued basic science understanding of the materials structure-property-processing relationship as it relates to AM will be key to mitigate risk for these industrial processes.

Acknowledgements The work reported here was a collaborative effort, and as such acknowledgement is due to others that contributed to the experimental work and/or critical discussions outside of those already called out in the text. This includes researchers at Purdue (Dr. Matthew Michie and Dr. Andres Diaz-Cano), Georgia Institute of Technology (Abel Diaz and Prof. Surya Kalidindi), and Johns Hopkins Applied Physics Laboratory (Dr. Brendan Croom). The views expressed here are those of the authors and do not reflect the official policy or position of the US Air Force, Department of Defense, or the US Government. Distribution A: Cleared for Public Release, #AFRL-2021-0114.

References

1. Padture, N.P.: Advanced structural ceramics in aerospace propulsion. Nat. Mater. **15**, 804–809 (2016). https://doi.org/10.1038/nmat4687
2. Allen, A.J., Levin, I., Witt, S.E.: Materials research & measurement needs for ceramics additive manufacturing. J. Am. Ceram. Soc. **103**, 6055–6069 (2020). https://doi.org/10.1111/jace.17369

3. Yang, L., Miyanaji, H.: Ceramic additive manufacturing: a review of current status and challenges. Solid Freeform Fabr., 652–679 (2017)
4. Chen, Z., Li, Z., Li, J., et al.: 3D printing of ceramics: a review. J. Eur. Ceram. Soc. **39**, 661–687 (2019). https://doi.org/10.1016/j.jeurceramsoc.2018.11.013
5. Wang, J.-C., Dommati, H., Hsieh, S.-J.: Review of additive manufacturing methods for high-performance ceramic materials. Int. J. Adv. Manuf. Technol. **103**, 2627–2647 (2019). https://doi.org/10.1007/s00170-019-03669-3
6. Zocca, A., Colombo, P., Gomes, C.M., Günster, J.: Additive manufacturing of ceramics: issues, potentialities, and opportunities. J. Am. Ceram. Soc. **98**, 1983–2001 (2015). https://doi.org/10.1111/jace.13700
7. Costa, E.C.E., Duarte, J.P., Bártolo, P.: A review of additive manufacturing for ceramic production. Rapid Prototyp. J. **23**, 954–963 (2017). https://doi.org/10.1108/RPJ-09-2015-0128
8. Zou, Y., Li, C.-H., Hu, L., et al.: Effects of short carbon fiber on the macro-properties, mechanical performance and microstructure of SiSiC composite fabricated by selective laser sintering. Ceram. Int. **46**, 12102–12110 (2020). https://doi.org/10.1016/j.ceramint.2020.01.255
9. Meyers, S., De Leersnijder, L., Vleugels, J., Kruth, J.-P.: Increasing the silicon carbide content in laser sintered reaction bonded silicon carbide. Ceram. Trans. **201**, 207–215 (2018)
10. Yi, X., Tan, Z.-J., Yu, W.-J., et al.: Three dimensional printing of carbon/carbon composites by selective laser sintering. Carbon N Y. **96**, 603–607 (2016). https://doi.org/10.1016/j.carbon.2015.09.110
11. King, D., Middendorf, J., Cissel, K., et al.: Selective laser melting for the preparation of an ultra-high temperature ceramic coating. Ceram. Int. **45**, 2466–2473 (2019). https://doi.org/10.1016/j.ceramint.2018.10.173
12. Costakis, W.J., Rueschhoff, L.M., Diaz-Cano, A.I., et al.: Additive manufacturing of boron carbide via continuous filament direct ink writing of aqueous ceramic suspensions. J. Eur. Ceram. Soc. **36**, 3249–3256 (2016). https://doi.org/10.1016/j.jeurceramsoc.2016.06.002
13. Rueschhoff, L., Costakis, W., Michie, M., et al.: Additive manufacturing of dense ceramic parts via direct ink writing of aqueous alumina suspensions. Int. J. Appl. Ceram. Technol. **13**, 821–830 (2016)
14. Kemp, J.W., Hmeidat, N.S., Compton, B.G.: Boron nitride-reinforced polysilazane-derived ceramic composites via direct-ink writing. J. Am. Ceram. Soc. **103**, 4043–4050 (2020). https://doi.org/10.1111/jace.17084
15. Zocca, A., Franchin, G., Elsayed, H., et al.: Direct ink writing of a preceramic polymer and fillers to produce hardystonite (Ca2ZnSi2O7) bioceramic scaffolds. J. Am. Ceram. Soc. **99**, 1960–1967 (2016). https://doi.org/10.1111/jace.14213
16. Franchin, G., Wahl, L., Colombo, P.: Direct ink writing of ceramic matrix composite structures. J. Am. Ceram. Soc. **100**, 4397–4401 (2017). https://doi.org/10.1111/jace.15045
17. Franchin, G., Maden, H., Wahl, L., et al.: Optimization and characterization of preceramic inks for direct ink writing of ceramic matrix composite structures. Materials (Basel). **11**, 515 (2018). https://doi.org/10.3390/ma11040515
18. Eckel, Z.C., Zhou, C., Martin, J.H., et al.: Additive manufacturing of polymer-derived ceramics. Science (80–). **351**, 58–62 (2016). https://doi.org/10.1126/science.aad2688
19. Zanchetta, E., Cattaldo, M., Franchin, G., et al.: Stereolithography of SiOC ceramic microcomponents. Adv. Mater. **28**, 370–376 (2016). https://doi.org/10.1002/adma.201503470
20. Wozniak, M., de Hazan, Y., Graule, T., Kata, D.: Rheology of UV curable colloidal silica dispersions for rapid prototyping applications. J. Eur. Ceram. Soc. **31**, 2221–2229 (2011). https://doi.org/10.1016/j.jeurceramsoc.2011.05.004
21. Halloran, J.W.: Ceramic stereolithography: additive manufacturing for ceramics by photopolymerization. Annu. Rev. Mater. Res. **46**, 19–40 (2016). https://doi.org/10.1146/annurev-matsci-070115-031841
22. Lewis, J.A.: Direct-write assembly of ceramics from colloidal inks. Curr. Opin. Solid State Mater. Sci. **6**, 245–250 (2002). https://doi.org/10.1016/S1359-0286(02)00031-1

23. Kemp, J.W., Diaz, A., Malek, E., et al.: Processing of ultra-high temperature ceramic matrix composites via direct-ink writing. In Review (2020)
24. Zhao, Z., Zhou, G., Yang, Z., et al.: Direct ink writing of continuous SiO2 fiber reinforced wave-transparent ceramics. J. Adv. Ceram. **9**, 403–412 (2020). https://doi.org/10.1007/s40145-020-0380-y
25. Lorenz, M., Dietemann, B., Wahl, L., et al.: Influence of platelet content on the fabrication of colloidal gels for robocasting: experimental analysis and numerical simulation. J. Eur. Ceram. Soc. **40**, 811–825 (2020). https://doi.org/10.1016/j.jeurceramsoc.2019.10.044
26. Muth, J.T., Dixon, P.G., Woish, L., et al.: Architected cellular ceramics with tailored stiffness via direct foam writing. Proc. Natl. Acad. Sci. **114**, 1832–1837 (2017). https://doi.org/10.1073/pnas.1616769114
27. Lewis, J.A., Smay, J.E., Stuecker, J., Cesarano III, J.: Direct ink writing of three-dimensional ceramic structures. J. Am. Ceram. Soc. **89**, 3599–3609 (2006). https://doi.org/10.1111/j.1551-2916.2006.01382.x
28. Colombo, P., Mera, G., Riedel, R., et al.: Polymer-derived ceramics: 40 years of research and innovation in advanced ceramics. J. Am. Ceram. Soc. **93**, 1805–1837 (2010). https://doi.org/10.1111/j.1551-2916.2010.03876.x
29. Baldwin, L.A., Rueschhoff, L.M., Deneault, J.R., et al.: Synthesis of a two-component carbosilane system for the advanced manufacturing of polymer-derived ceramics. Chem. Mater. **30**, 7527–7534 (2018). https://doi.org/10.1021/acs.chemmater.8b02541
30. Youngblood, J.P., Trice, R.W., Wiesner, V.L., et al.: Injection molding of aqueous suspensions of high-temperature ceramics. US Patent Application 15/192,376, filed 6/2016, published 5/2017, 2015
31. Wiesner, V.L., Youngblood, J.P., Trice, R.W.: Room-temperature injection molding of aqueous alumina-polyvinylpyrrolidone suspensions. J. Eur. Ceram. Soc. **34**, 453–463 (2014). https://doi.org/10.1016/j.jeurceramsoc.2013.08.017
32. Acosta, M., Wiesner, V.L., Martinez, C.J., et al.: Effect of polyvinylpyrrolidone additions on the rheology of aqueous, highly loaded alumina suspensions. J. Am. Ceram. Soc. **96**, 1372–1382 (2013). https://doi.org/10.1111/jace.12277
33. Riley, F.L.: Silicon nitride and related materials. J. Am. Ceram. Soc. **83**, 245–265 (2000)
34. Mei, H., Yan, Y., Feng, L., et al.: First printing of continuous fibers into ceramics. J. Am. Ceram. Soc. **102**, 3244–3255 (2019). https://doi.org/10.1111/jace.16234
35. Rueschhoff, L.M., Carney, C.M., Apostolov, Z.D., Cinibulk, M.K.: Processing of fiber-reinforced ultra-high temperature ceramic composites: a review. Int. J. Ceram. Eng. Sci. **2**, 22–37 (2020). https://doi.org/10.1002/ces2.10033
36. Sciti, D., Guicciardi, S., Silvestroni, L.: SiC chopped fibers reinforced ZrB2: effect of the sintering aid. Scr. Mater. **64**, 769–772 (2011). https://doi.org/10.1016/j.scriptamat.2010.12.044
37. Guicciardi, S., Silvestroni, L., Nygren, M., Sciti, D.: Microstructure and toughening mechanisms in spark plasma-sintered ZrB 2 ceramics reinforced by SiC whiskers or SiC-chopped fibers. J. Am. Ceram. Soc. **93**, 2384–2391 (2010). https://doi.org/10.1111/j.1551-2916.2010.03730.x
38. Fang, C., Hu, P., Dong, S., et al.: Design and optimization of the coating thickness on chopped carbon fibers and sintering temperature for ZrB2-SiC-Cf composites prepared by hot pressing. J. Eur. Ceram. Soc. **39**, 2805–2811 (2019). https://doi.org/10.1016/j.jeurceramsoc.2019.03.038
39. Wiesner, V., Acosta, M., Rueschhoff, L., et al.: Horizontal Dip-Spin Casting of aqueous alumina-polyvinylpyrrolidone suspensions with chopped fiber. Int. J. Appl. Ceram. Technol. **14**, 1077–1087 (2017). https://doi.org/10.1111/ijac.12714
40. Silvestroni, L., Capiani, C., Dalle Fabbriche, D., Melandri, C.: Novel light and tough ZrB2-based functionally graded ceramics. Compos. Part B Eng. **99**, 321–329 (2016). https://doi.org/10.1016/j.compositesb.2016.06.001
41. Compton, B.G., Lewis, J.A.: 3D-printing of lightweight cellular composites. Adv. Mater. **26**, 5930–5935 (2014). https://doi.org/10.1002/adma.201401804

42. Van Wie, D.M., Drewry Jr., D.G., King, D.E., Hudson, C.M.: The hypersonic environment: required operating conditions and design challenges. J. Mater. Sci. **39**, 5915–5924 (2004). https://doi.org/10.1023/B:JMSC.0000041688.68135.8b

43. Silvestroni, L., Sciti, D., Melandri, C., Guicciardi, S.: Toughened ZrB2-based ceramics through SiC whisker or SiC chopped fiber additions. J. Eur. Ceram. Soc. **30**, 2155–2164 (2010). https://doi.org/10.1016/j.jeurceramsoc.2009.11.012

44. Fahrenholtz, W.G., Hilmas, G.E.: Ultra-high temperature ceramics: materials for extreme environments. Scr. Mater. **129**, 94–99 (2017). https://doi.org/10.1016/j.scriptamat.2016.10.018

45. Alfano, D., Gardi, R., Scatteia, L., Del Vecchio, A.: UHTC-based hot structures: characterization, design, and on-ground/in-flight testing. In: Fahrenholtz, W.G., Wuchina, E., Lee, W.E., Zhou, Y. (eds.) Ultra-High Temperature Ceramics: Materials for Extreme Environment Applications, pp. 416–436. Wiley, New York (2014)

46. Sciti, D., Pienti, L., Fabbriche, D.D., et al.: Combined effect of SiC chopped fibers and SiC whiskers on the toughening of ZrB2. Ceram. Int. **40**, 4819–4826 (2014). https://doi.org/10.1016/j.ceramint.2013.09.031

47. Sha, J.J., Li, J., Wang, S.H., et al.: Improved microstructure and fracture properties of short carbon fiber-toughened ZrB2-based UHTC composites via colloidal process. Int. J. Refract. Met. Hard Mater. **60**, 68–74 (2016). https://doi.org/10.1016/j.ijrmhm.2016.07.010

48. Leslie, C.J., Boakye, E.E., Keller, K.A., Cinibulk, M.K.: Development of continuous SiC fiber reinforced HfB2-SiC composites for aerospace applications. In: Processing and Properties of Advanced Ceramcis and Composites V: Ceramic Transactions, pp. 3–12. John Wiley & Sons, Inc., Hoboken (2013)

49. Leslie, C.J., Kim, H.J., Chen, H., et al.: Polymer-derived ceramics for development of ultra-high temperature composites. In: Innovative Processing and Manufacturing of Advanced Ceramics and Composites II, pp. 33–46. John Wiley & Sons, Inc., Hoboken (2014)

50. Silvestroni, L., Sciti, D., Hilmas, G.E., et al.: Effect of a weak fiber interface coating in ZrB2 reinforced with long SiC fibers. Mater. Des. **88**, 610–618 (2015). https://doi.org/10.1016/j.matdes.2015.08.105

51. Yang, F., Zhang, X., Han, J., Du, S.: Characterization of hot-pressed short carbon fiber reinforced ZrB2–SiC ultra-high temperature ceramic composites. J. Alloys Compd. **472**, 395–399 (2009). https://doi.org/10.1016/j.jallcom.2008.04.092

52. Carney, C.M.: Ultra-high temperature ceramic-based composites. In: Zweben, C.H., Beaumont, W.R. (eds.) Comprehensive Composite Materials II, vol. 5, pp. 281–292. Elsevier, Amsterdam (2017)

53. Corral, E.L., Walker, L.S.: Improved ablation resistance of C-C composites using zirconium diboride and boron carbide. J. Eur. Ceram. Soc. **30**, 2357–2364 (2010). https://doi.org/10.1016/j.jeurceramsoc.2010.02.025

54. Tang, S., Hu, C.: Design, preparation and properties of carbon fiber reinforced ultra-high temperature ceramic composites for aerospace applications: a review. J. Mater. Sci. Technol. **33**, 117–130 (2017). https://doi.org/10.1016/j.jmst.2016.08.004

55. Key, T.S., Wilks, G.B., Parthasarathy, T.A., et al.: Process modeling of the low-temperature evolution and yield of polycarbosilanes for ceramic matrix composites. J. Am. Ceram. Soc. **101**, 2809–2818 (2018). https://doi.org/10.1111/jace.15463

56. Croom, B., Abbott, A., Kemp, J.W., et al.: Mechanics of nozzle clogging during direct ink writing of fiber-reinforced composites. Addit. Manuf. **37**, 101701 (2021)

57. Folgar, F., Tucker, C.L.: Orientation behavior of fibers in concentrated suspensions. J. Reinf. Plast. Compos. **3**, 98–119 (1984). https://doi.org/10.1177/073168448400300201

58. Apostolov, Z.D., Heckman, E.P., Key, T.S., Cinibulk, M.K.: Effects of low-temperature treatment on the properties of commercial preceramic polymers. J. Eur. Ceram. Soc. **40**, 2887–2895 (2020). https://doi.org/10.1016/j.jeurceramsoc.2020.02.030

59. Pelz, J.S., Ku, N., Shoulders, W.T., Meyers, M.A., Vargas-Gonzalez, L.R.: Multi-material additive manufacturing of functionally graded carbide ceramics via active, in-line mixing, Addit. Manuf. (2020) 101647. https://doi.org/10.1016/j.addma.2020.101647

Dr. Lisa Rueschhoff is a materials research engineer in the Composites Branch at the Air Force Research Laboratory (AFRL) in Wright Patterson Air Force Base, Ohio. She develops and leads in-house research in the Ceramic Materials & Processes Team in the area of high-temperature structural ceramics and composites using traditional processing as well as additive manufacturing techniques. Her work aims to address existing performance and processing issues in ceramics to enable next-generation aerospace materials. She has been involved in materials processing research since her work at Ames National Laboratory as a student researcher while obtaining her B.S in materials engineering from Iowa State University. Her research there focused on metal alloy development for lithium-ion batteries. While at Iowa State, she was an active member of the Material Advantage organization, serving as chapter chair and earning the Most Outstanding Chapter in the Nation Award in 2013. Other awards from Iowa State include Exemplary Peer Mentor and the MSE Student Leadership and Service Award. Her research focus shifted to ceramic processing, and specifically additive manufacturing of ceramics, during her Ph.D. research at Purdue University. There she earned a NSF Graduate Research Fellowship to develop novel processing routes for creating complex-shaped advanced ceramics. Her research resulted in three first-author manuscripts and 15 research presentations at conferences and national laboratories. She was active in the Purdue MSE Graduate Research Society, received the College of Engineering Magoon Award for Excellence in Teaching and the MSE Outstanding Graduating Graduate Student Award. She joined AFRL in 2017, first as a National Research Council (NRC) Postdoctoral Fellow and then as a government civilian member of the team in 2018. As part of her role, she mentors students and post-doctoral researchers in ceramic additive manufacturing for aerospace applications. She is actively involved in the ceramics community through participation and leadership in the American Ceramic Society (ACerS). She serves on various ACerS committees, is an associate editor of the *International Journal of Applied Ceramic Technology*, and co-organizes events and symposia at their research conferences. For this work in the society and community, she was awarded the ACerS 2019 International Jubilee Global Diversity Award and an ACerS Global Ambassador Award in 2017. She was an invited participant in the 2019 National Academies of Engineering US Frontiers of Engineering Symposia where she participated in cross-disciplinary discussions with the goal of building and improving US innovative capacity.

Chapter 11
The Additive Journey from Powder to Part

Yanli Zhu, Ahmet Okyay, Mihaela Vlasea, Kaan Erkorkmaz,
and Mark Kirby

Assembling a Talented Team to Solve a Complex Multidisciplinary Problem

Additive manufacturing (AM) has received growing attention from the industry. This is partially attributed to the fact that shape complexity can be achieved with ease, whereas conventional manufacturing methods have restrictions in terms of fabrication of geometries. As such, owing to the tool-less nature of the process, AM has advantages in small batch production of high-value-added customized components. Laser powder bed fusion (LPBF), a type of AM, utilizes a laser beam to successively melt layers of metal powders to form a near-net shape part. The quality of AM parts is sensitive to print parameters and materials of choice. Inherently, poor surface finish, geometric distortion due to thermal-induced residual stress, and anisotropic properties are commonly observed in AM parts. As a result, in order to produce complex parts with high-precision features such as mating surfaces and threaded holes, it is desired to post-machine critical features on an AM part to achieve the required dimensional tolerance. Subtractive manufacturing (SM) approaches will thus also need to evolve in tandem to respond to finishing needs for complex geometries.

In order to holistically utilize the capabilities of AM and SM, hybrid manufacturing solutions have been proposed in literature and in industry, which can be mainly categorized into an integrated process alternating between AM and SM or a separated AM process followed by SM. Therefore, when a high degree of design complexity is involved, a diverse skillset is required to go from powder to product. In

Y. Zhu · A. Okyay · M. Vlasea (✉) · K. Erkorkmaz
University of Waterloo, Waterloo, ON, Canada
e-mail: mihaela.vlasea@uwaterloo.ca; y68zhu@uwaterloo.ca; ahmet.okyay@uwaterloo.ca; kaane@uwaterloo.ca

M. Kirby
Renishaw, Wotton-under-Edge, UK
e-mail: mark.kirby@uwaterloo.ca

© Springer Nature Switzerland AG 2021
S. M. DelVecchio (ed.), *Women in 3D Printing*, Women in Engineering and Science, https://doi.org/10.1007/978-3-030-70736-1_11

deploying AM, the knowledge of the complex relationships between material-process-performance is crucial in producing sound parts with good mechanical properties. In addition, AM design-specific requirements, such as but not limited to overhang regions requiring support structures, feature size, orientation-dependent geometric qualities, and part distortion, demand additional skillsets. Challenges of post-machining of AM products include machine tool accessibility, machining allowance and trajectories planning, datum selection, fixturing methods for complex near-net-shape parts, vibration minimization under cutting forces, variable properties, and many more.

As a result, thorough process planning for the synergized AM and SM process is essential in order to minimize trial and error attempts and scrap parts. This chapter focuses on the story behind bringing to life an extremely complex surgical instrument by consolidating the right team with the right skillsets to make it happen. Such a team was assembled by Dr. Vlasea, including Dr. Vlasea and her student Yanli Zhu (additive manufacturing expertise), including Dr. Erkorkmaz and Dr. Okyay (machining and controls expertise), as well as the passionate industry partner Renishaw Solutions Centre with direct input from Mark Kirby and our industry collaborator Intellijoint Surgical. This chapter describes the technical efforts behind the journey from powder to part, through failure and success alike.

Need for Synergy in Additive Manufacturing and Machining Strategies

Additive manufacturing (AM) is recognized for its potential for complementing conventional manufacturing methods such as subtractive, formative, and joining processes. One of the promising features inherent to the nature of AM technologies is the design freedom and the potential for consolidating assemblies [1–3], the integration of lattice structures for light weight [4–6], the realization of topology optimization for high strength-to-weight-ratio components [7–11], and the customization of part design [12–14]. These benefits are carefully considered against cost factors when deciding on the most appropriate manufacturing method. Depending on the product, there is often a breakeven point where the benefit in design complexity and/or time-to-market justifies the choice. In this chapter, AM design rules are considered from the perspective of laser powder bed fusion (LPBF) and subtractive machining (SM) for complex geometries.

Recently, the development of AM technologies has enabled the small batch production of customized biomedical implants and surgical tooling with added value. Tuomi et al. classified the AM application in the medical field into five major areas: (1) medical models, (2) surgical implants, (3) surgical guides, (4) external aids, and (5) bio-manufacturing [15]. With AM and 3D computed tomography (CT) technologies, more precise and customized implants can be designed and fabricated cost-effectively [16]. Porous metal bone scaffolds and orthopedic implants manufactured via AM are suitable for replacing or reconstructing damaged bones since their stiffness and porosity can be customized as required to allow for bone tissue fixation [17, 18]. AM technologies have also been applied to surgical tooling due to the capability to manufacture customized patient-specific tools on demand. As an example, researchers at the Brigham Young University were able to miniaturize pin joints of a suturing device by using AM to produce a smaller yet more precise surgical instrument [19]. As another example, researchers manufactured a more flexible and accurate tool used in an anterior cruciate ligament (ACL) reconstruction surgery by utilizing LPBF of Inconel 718 alloy [20]. There are many more examples of surgical tools fabricated using AM are emerging, showcasing the design and fabrication possibilities of this technology [21, 22].

Despite the advantages and potentials of design for AM, there are inherent drawbacks that hinder the wider adoption of AM technologies. Compared to SM methods, current metal AM methods suffer from poor surface finish and part dimensional inaccuracies [23]. The typical order of magnitude of the surface roughness (S_a: arithmetical mean height of a surface) of a milling surface is between 10^{-1} and $10^0 \mu m$ [24], while that of a LPBF surface is between 10^0 and 10^1 μm, depending on geometry and process parameters [25–27]. Another challenge in AM is the presence of internal and sub-surface pores which are prone to occur. These types of pores affect mechanical properties such as fatigue life and tensile performance [28–31]. In addition, AM parts usually exhibit anisotropic mechanical properties with the build direction generally being the weakest [32, 33]. Due to the repeated, rapid, and concentrated thermal cycling, residual stresses are accumulated in AM parts. These in turn lead to geometric distortion and dimensional inaccuracies, following part removal from the build plate [34–36]. For the purpose of part anchoring, weight supporting, and heat dissipation, support structures are usually required for LPBF parts with overhangs of angles smaller than 45° with respect to the build plate [37, 38]. For AISI 316 L, an overhang parallel to the build plate larger than 1.5 mm length is not possible to produce without a support structure [39]. For aluminum alloy the maximum parallel overhang length is 2 mm [40]. Post-processing is necessary to attain the desired functionality in parts. Post-processing methods such as heat treatment and hot isostatic pressing (HIP) are required to reduce residual stress and improve part homogeneity [41–43]. As-built surfaces are also usually finished by sandblasting, electro-polishing, machining, laser etching, etc., there may also be a need for the secondary machining of high precision features such as mating surfaces, holes, and treads, as well as overhang surfaces which require support structures.

Complex, lightweight parts can be fabricated via AM, while tolerance requirements can be fulfilled by SM, the two manufacturing technologies complementing each other. There are mainly two existing hybrid approaches for AM and SM. One approach is the integrated, alternating AM and SM processes, which usually integrate a milling tool in a powder bed/fed AM system [44, 45]. Milling is performed intermittently, after the fabrication of several layers. An advantage of this approach is that the effect of surface defects on successive layers can be minimized. Also, part fixation and datum alignment are usually unnecessary as the part is welded on the build plate. However, coolant is prohibited in the integrated hybrid systems, which may lead to reduced tool life. In addition, the part may still deform out of tolerance after removal from the build plate. The other approach is to have separated successive processes which allow for the production of different batches in the AM and SM machines simultaneously [23, 46]. As a result, manufacturing time can be reduced. Coolant is not a concern for the separated processes. This chapter focuses on highlighting efforts in process planning for this latter approach for a component with a high degree of design complexity.

Challenges for the fabrication of complex near-net-shape via AM, followed by post-machining, are not only inherent to the AM and CNC machining processes independently, but also spring from the overlapping performance criteria of the combined processes. For instance, industry case studies reported part vibration and fracture issues when machining lightweight AM structures, due to the reduction in stiffness [47]. It was also identified as a challenge to clamp a complex AM part in the CNC machine-tool. To accommodate this, a 3D plastic encapsulating fixture was printed to mount an AM microwave guide on the CNC machine for surface finishing [47]. Oyelola et al. investigated the machining behavior and surface integrity of Ti-6Al-4 V components produced by direct metal deposition. Results showed that the inhomogeneity in microstructure has effects on the cutting forces and chip formation during machining [47, 48]. Frank et al., in the description of a direct additive-subtractive hybrid (DASH) manufacturing method, emphasized the importance of identifying the features to be machined and their corresponding tolerance. Based on the identified features and cutter accessibility, plans for part orientation, clamping, and machining allowance should be made in the design stage [49].

The purpose of this work is to generate a workflow for combining AM and post-machining for highly complex geometries achievable via LPBF. The following objectives are of interest in this study: CAD design using topology optimization, AM process planning, machining process planning, fabrication, and validation. The goal is to be able to deploy and iterate on the workflow through a case study of manufacturing a surgical tool using LPBF followed by SM, in order to produce a part which fulfills all the functional and dimensional specifications.

Case Study for a Synergistic Approach in Additive Manufacturing and Machining

Surgical Navigation Tracker Demonstrator Design Criteria

In this work, a surgical navigation tracker (Fig. 11.1) is presented as a case study for redesign and fabrication via AM, followed by SM of critical surfaces. During hip replacement surgeries, the tracker is attached to a probing rod by a kinematic mount.

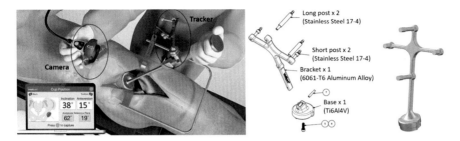

Fig. 11.1 Original design of the surgical navigation tracker. (Courtesy of Intellijoint Surgical Inc.)

Table 11.1 Tracker design criteria

Criterion no.	Functional design criteria description
1	The relative position of the posts and the base must be within ±25 μm
2	It is desired that the tracker will not impinge on soft tissues of the patient
3	It should be graspable by an adult hand without touching the tips of the posts
4	The tracker body should be non-reflective with a matte surface
5	All spheres attached to the posts should be visible to the camera
6	The center of gravity should be as low as possible to avoid toppling
7	The overall height of the tracker should not exceed 150 mm
8	External mating parts interfacing with this tracker should not be modified

While probing on several locations on the patient's leg, a camera mounted on the patient's pelvis takes images of the reflective spheres mounted on the tracker in order to calculate the cup position, leg length, and offset. The original tracker is an assembly consisting of a Ti-6Al-4 V (Ti64) base, an Al 6061-T6 bracket, a pin, a screw, and four stainless steel (SS) 17–4 posts at the tips of which the reflective spheres are press-fitted. The most important constraint of the design is that the dimensional tolerance of the location of the spheres referenced to the bottom surface of the base should be within ±25 μm. Other design criteria can be referred to in Table 11.1.

In the present assembly, the specified tolerance is difficult to achieve, due to the accumulation of dimensional uncertainty of press-fitting in the bracket and posts. With AM, it is possible to consolidate the original design and print the whole part using Ti64. A high strength-to-weight-ratio design can be obtained using structural topology optimization. The optimized complex structure can be realized by LPBF. In parallel, manufacturing constraints imposed by the AM process such as part orientation and overhangs need to be considered and will be described in detail in the section on the design of the workholding fixtures.

Digital Workflow for AM and Post-Machining

The surgical tracker design presents a challenge and opportunity to study the synergy between AM topology optimization potential and secondary processing of complex structures via SM. The team proposed a workflow strategy for the integration of AM and CNC post-machining, as shown in Fig. 11.2. The process of manufacturing a successful finished part flows through the six modules iteratively.

A. Digital design environment – AM design potentials are developed.
B. AM build environment –– Build capabilities to address challenges such as minimizing material use and addressing overhang constraints are analyzed.
C. AM simulation and/or verification – Predicted geometric distortion is considered.
D. Experimental domain – The designs are re-iterated, including generating new features and the fixturing and/or clamping necessary to address the needs for stabilizing the structure during the AM (distortion) and SM (vibration) process by quantifying the AM material response to SM cutting forces.
E. Post-machining planning – Tool accessibility and minimal vibration are ensured.

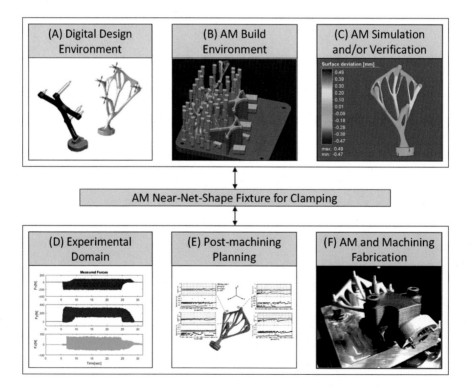

Fig. 11.2 AM and SM process workflow, illustrating a six-step iterative workflow, where designs are conceptualized and optimized

F. AM and machining fabrication – Parts are fabricated sequentially via AM and SM.

AM thermal distortion simulations for complex topologically optimized shapes, as well as SM experimental datasets and simulations, are essential for the effective planning of the synergized manufacturing process. By following a methodical approach, the costly iteration in experiments and part scrap during the post-processing step can be minimized. Distortion simulation is not the focus of this study; however, the resulting geometric distortion of AM parts was measured using a 3D optical scanner. In addition to these simulations and measurements, complex clamping methods for machining are compared in terms of vibration minimization and machining stability. This work is intended to showcase design optimization in accordance with the potentials and capabilities of both AM manufacturability and SM machinability.

Design of the Surgical Tracker Via Topology Optimization (Fig. 11.2 – Workflow A)

SolidThinking Inspire® (hereafter referred to as Inspire), which implements the SIMP (Solid Isotropic Material with Penalty) topology optimization algorithm, was used to generate the design candidates. As it is shown in Fig. 11.3, the part in brown was defined as the design space and the post tips and tacker base (in silver) were defined as the non-design space. The design space was created with the consideration of functional criteria of the tracker (Table 11.1).

In order to create a structure in which the posts and the base are rigidly connected with each other, the combination of different load cases was applied in the topology optimization step, as shown in Fig. 11.3f. The magnitude of the resulting force acting on each post is 30 N. Based on user input, this force is approximately three times the maximum force that the product will be expected to see during the press-fitting

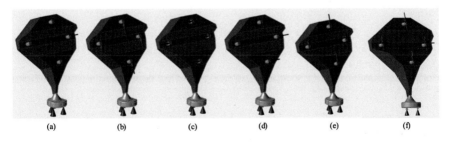

(a)   (b)   (c)   (d)   (e)   (f)

Fig. 11.3 (**a**) Load case 1 with normal acting forces; (**b**) load case 2 with upwards acting forces; (**c**) load case 3 with rightwards acting forces; (**d**) load case 4 with leftwards acting forces; (**e**) load case 5 with downwards acting forces; (**f**) multi-directional loads

Table 11.2 Topology optimization settings

Load direction	Load magnitude	Volume fraction	Overhang constraint
Multiple	30 N	30%, 25%, 20%, 15%, 10%, 5%	N/A

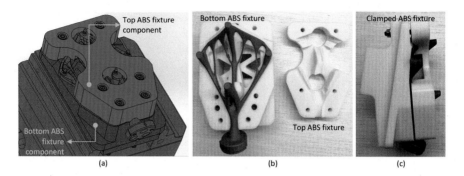

Fig. 11.4 The ABS encapsulating fixture (**a**) illustrating the design of the top and bottom fixture components, (**b**) the manufactured ABS fixture components, and (**c**) the encapsulated tracker inside the clamped ABS fixture

of the reflective spheres. The base of the tracker is always fixed. Table 11.2 gives a summary of the settings used in the topology optimization step.

Design of the Workholding Fixtures (Fig. 11.2 – Workflow A)

Due to the highly complex geometrical nature of the parts producible by AM, specialized fixtures are required to hold the workpiece in the machine-tool vise. Two fixture concepts were explored. Firstly, an encapsulating fixture was designed and additively manufactured in two pieces using acrylonitrile butadiene styrene (ABS) material in a Stratasys F350 printer (Fig. 11.4). The top and bottom ABS workholding pieces (Fig. 11.4a, b) were designed to encapsulate the tracker. The cavity of each top and bottom fixture was created by subtracting the CAD model of the tracker resulting from the topology optimization step (Fig. 11.4b). The topology optimization results are presented later in this chapter. While designing the workholding fixture, it has been ensured that all surfaces to be machined were reachable by the cutting tool during machining. This workholding solution results in clamping forces which may distort the tracker component.

Alternatively, a novel workholding system by Blue Photon® was used to stabilize the tracker directly under the tracker posts with UV cured adhesive polymer (Fig. 11.5a, b). The Blue Photon® workholding consists of fixture posts Fig. 11.5b with central conduits for guiding UV light onto a UV-curable polymer. The UV curable polymer is in contact with both the fixture post and the part, thus fixing the part upon UV crosslinking. The polymer can be manually removed after machining is completed. This workholding solution is suitable for the fixation of a flexible

Fig. 11.5 The UV fixture (**a**) with the tracker mounted on top of the UV Fixture posts showing the tracker areas which cannot be supported, (**b**) a close-up of the UV fixture post and UV-curable adhesive polymer, and the (**c**) ABS secondary add-on fixture

complex-shape workpiece, while avoiding the clamping-induced distortion from the possible over-tightening of a fully encapsulating fixture. However, a limitation exists as to the accessibility of the UV device to some regions of the part. For example, the UV fixture posts are blocked from the lower post of the tracker as seen in Fig. 11.5a. To avoid severe vibration when machining the lower post, since this area is not mechanically constrained, a small ABS fixture produced via AM and encapsulating the lower post was added as illustrated in Fig. 11.5c.

Additive Manufacturing of the Tracker Design and Machining Artifacts (Fig. 11.2 – Workflow B)

The print orientation constraint was chosen with the base of the tracker anchored onto the build plate by supports. The selected orientation not only reduces the unit cost of parts by accommodating more trackers on one build plate, but also minimizes thermally induced geometric distortions by avoiding large cross-section variability between print layers. Structurally optimized parts may contain overhangs which require support structures. As a result, the structural topology design was manually modified to create a set of self-supporting part families which do not require support structures or subsequent post-processing in terms of support removal.

AM artifacts of dimension 20×30×40 mm were designed and printed along with the trackers for the purpose of identifying cutting force coefficients which are crucial for cutting force prediction and chatter stability analysis. The trackers were printed using stripe scanning strategy. The AM artifacts for cutting tests were printed using both stripe and meander scanning strategies to study the effect of scanning strategy on the cutting force coefficients. The meander scanning strategy infills the layer in a straight line vector from each side of the part border, whereas the stripe scanning strategy splits the area within the part perimeter into strips, where each strip is printed using a meander strategy. The trackers and the artifacts were printed using Ti-6Al-4 V ELI-0406 (Ti64), a Titanium alloy powder provided by Renishaw®, on their AM 400 machine.

Table 11.3 QuantAM print parameters for Ti64 with stripe pattern

QuantAM – stripe

	Power (W)	Beam width (mm)	Exposure time (ms)	Hatch point distance (mm)	Hatch distance (mm)	Stripe size (mm)	Stripe offset (mm)	Layer thickness (mm)
Volume	200	0.075	50	55	0.105	5	0.01	0.03
Border	100	0.075	40	45	–	–	–	–
Upskin	175	0.075	75	50	0.065	–	–	0.03
Downskin	175	0.075	75	50	0.065	–	–	0.03

Table 11.4 QuantAM print parameters for Ti64 with meander pattern

QuantAM – meander

	Power (W)	Beam width (mm)	Exposure time (us)	Hatch point distance (mm)	Hatch distance (mm)	Layer thickness (mm)
Volume	200	0.075	50	75	0.065	0.03
Border	100	0.075	40	45	–	–
Upskin	175	0.075	50	75	0.065	0.03
Downskin	175	0.075	50	75	0.065	0.03

The tracker and machining artifacts with stripe pattern were printed using the QuantAM (Renishaw®) recipe for Ti64 with stripe pattern as summarized in Table 11.3. The meander machining artifacts were printed using the QuantAM process parameters as shown in Table 11.4. The layer thickness used is 30 μm. Laser power and exposure time for the border is reduced compared to the volume to compensate for the low heat conduction rate of un-melt powder.

Dimensional Distortion Measurements (Fig. 11.2 – Workflow C)

The AICON® optical scan system, which was calibrated to an accuracy of 2 μm prior to scanning, was used to capture the 3D image of the printed tracker. The obtained point cloud data of the printed tracker was post-processed using the inspector module of Polyworks® to measure the geometric deviation of the scanned data from the CAD. The scanned data was best-fitted with the CAD at the tracker base, and distortion measurements were taken along the cross sections at the posts and at the base. The "best fit" alignment compared every point in the scanned data with their nearest neighboring point in the CAD and calculated a displacement distance. By applying the least squares algorithm, the scanned point set was transformed such that the sum of the squared distance between the matching points in the two sets is minimal.

Mechanical Vibration Identification (Fig. 11.2 – Workflow D & E)

Modal testing of the tracker with and without confinement to the proposed fixtures were carried out to determine the dynamic response of the posts. Results were compared to verify the stiffness contribution of the fixtures. Subsequent frequency domain chatter stability analyses were conducted to further illustrate differences in machining stability across different cases.

Modal Testing Using Impact Hammer

Impact hammer testing was applied to the tracker posts to identify the variation of the frequency response functions (FRF), i.e., the dynamic compliance, across different fixtures. The cutting tools were also tested, as it is required in the calculation of frequency domain chatter stability diagrams.

The final tracker design was tested using a Dytran 3035AG accelerometer, a Dytran 5800SL impact hammer, and LMS Test.Lab software. Four posts of the structure were excited by impacts applied from different directions, as shown in Fig. 11.6 with the results.

The 5-teeth Ø8 mm milling-tool (Sandvik) and the 4-teeth Ø12mm form-cutting tool (Tnt tool Inc.) were tested on the same 5-axis milling machine used in the machining experiments, using Dytran 3035 AG accelerometer and PCB 086C01 impact hammer. Both the feed (X) and normal (Y) directions were tested, as shown in Fig. 11.6a for the milling tool. Figure 11.6b is a picture of the form tool.

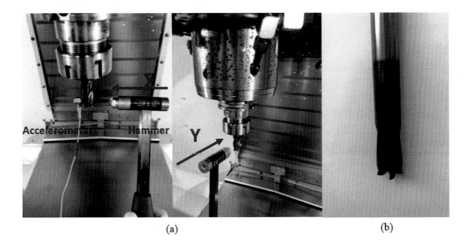

(a)   (b)

Fig. 11.6 (**a**) Impact hammer testing at milling tool tip in X and Y directions; (**b**) form tool

Frequency Domain Chatter Stability Analysis

For the frequency domain chatter stability analysis, FRFs obtained from the tools and the tracker were entered to CutPRO® (machining simulation software, MAL, Inc.) along with the cutting coefficients from cutting tests explained in the next section. Down-milling with 3/4 – immersion was assumed for both tools in each location of the machined post. The software makes use of the chatter stability theory by Budak [50] to calculate the stability lobe diagrams.

Cutting Tests and Machining (Fig. 11.2 – Workflow D & F)

For machining, the 5-axis milling machine (Matsuura MX-520) in the Additive Manufacturing Solutions Centre (Kitchener) of Renishaw Corp. was used. The posts and the base were machined subsequently, after clamping the fixtures on the vise. This way, the dimensional conformity of the machined details and their relative positions to each other were assured, regardless of possible deviations in the part directly obtained from AM. In this section, first, the cutting tests for the identification of the cutting coefficients (K_{tc}, K_{rc}, K_{te}, K_{re}) were presented. Finally, the machining parameters, including machine inputs and the cutting tools, were shown.

Tests for the Identification of the Cutting Coefficients

A series of machining tests with the Ti64 meander and stripe artifacts were conducted on the 3-axis Haas® VF-2YT milling-machine to identify the cutting coefficients. A 4-teeth Ø12.7 mm coated carbide helical end mill in down-milling was used with 1/2 – immersion, 1 mm depth of cut, and 1750 RPM spindle speed. The cutter was run at six different feed rates while keeping other cutting parameters constant. Each set of cutting tests was repeated 6 layers into the artifacts (Fig. 11.7).

Tracker Machining Parameters

The 5-teeth Ø8 mm milling tool was used to rough (Fig. 11.8a) and finish (Fig. 11.8b) the equator diameter of the post. A detailed view of the post is presented in Fig. 11.8c. The 4-teeth Ø12mm form tool was used to produce the chamfer form above and beneath the equator. In all cases, the tools were cutting at full axial depth in the targeted geometries. The machining parameters for the tracker with the ABS enclosure and the adhesive polymer fixture are presented for the machining of the posts in Table 11.5.

Fig. 11.7 Milling tests of the artifact

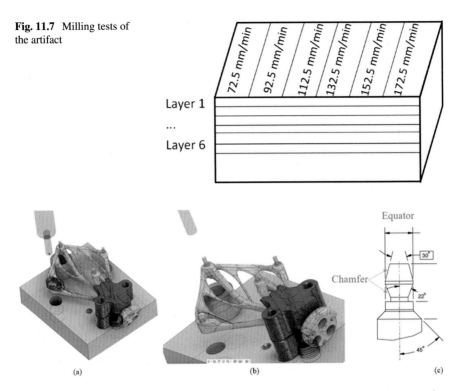

Fig. 11.8 Visualization of (**a**) roughing; (**b**) finishing of the equator; (**c**) finishing of the chamfer

Table 11.5 Machining parameters

Passes	Cutter	Radial immersion (mm)	Spindle speed (rpm)	Feed rate (mm/min)
Roughing	8 mm milling tool	0.2	4890	1930
Finishing of the equator	8 mm milling tool	0.1	6230	1650
Finishing of the chamfer	12.7 mm form tool	Variable (0 mm at equator)	2700	155

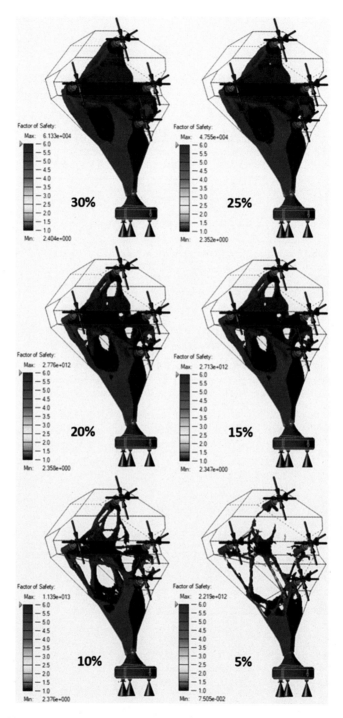

Fig. 11.9 Topology optimization designs with 30%, 25%, 20%, 15%, 10%, 5% volume constraints

Case Study Technical Outcomes, Reflections, and Iterations

Topology Optimization

A series of topology optimization designs were generated based on the design criteria listed in Table 11.1 and the topology optimization settings listed in Table 11.2. The design outcomes are shown in Fig. 11.9. To avoid distortion, the design space was selected such that the final posts may be interconnected. From this figure, it can be seen that at approximately 5% volume fraction, discontinuities start to occur in the design solution.

To assist in the selection of the most appropriate design based on the volume fraction versus performance, a Pareto front plot was generated (Fig. 11.10) showing the relationship between the design volume fraction and the compliance (strain energy). In this figure, it is shown that for the design space selected, when the volume fraction constraint is larger than 10%, the reduction in compliance stagnates. As the objectives of this topology optimization are to minimize the strain energy (i.e., minimize compliance) while keeping the design as light as possible, the design with 10% volume fraction was used as a reference to create a smooth self-supporting final design.

In Inspire, polyNURBS tool (Altair Inspire) was used around the topology structure to smooth out the facets, while keeping in mind that all overhangs should have angles larger than 45° relative to the build plate. In addition, the minimum features of the tracker were carefully considered to avoid any regions with a feature size smaller than 3 mm. A smooth and self-supporting tracker design was obtained as shown in Fig. 11.11. To ensure tolerance of the final tracker posts with respect to the tracker base, a machining allowance of 5 mm extra material was added to each one of the posts and the base of the design. This ensured there was enough material to remove in the machining process, despite the possible distortion during manufacturing.

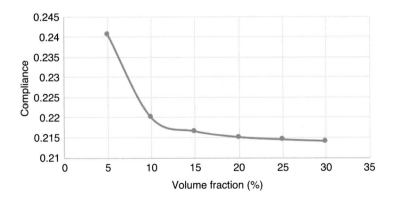

Fig. 11.10 Pareto front between compliance and volume fraction

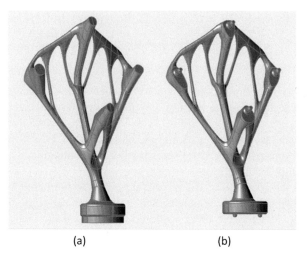

(a)   (b)

Fig. 11.11 Visualization of the tracker CAD design (**a**) before machining cuts are applied and (**b**) after machining cuts are applied

Table 11.6 Tracker design performance

Design	Total mass	Volume fraction	Compliance	Safety factor
Current, consolidated	76.5 g	9.7%	0.1967	3.178
Original, assembly	54 g	–	–	0.083

Stress analysis was performed on the final tracker design to evaluate its performance under the same loading conditions as applied in topology optimization and the results are provided in Table 11.6. The total mass of the consolidated tracker design is 73.5 g while the original tracker assembly illustrated in Fig. 11.1 is 54 g. With loadings of 30 N acting on the posts from multiple directions, the safety factor of the consolidated tracker is 3.178. Although the current design is heavier than the original design, it has a higher safety factor under the same loading conditions. A stiff part is also desired to prevent failure during post-machining. Hence, the current consolidated design was printed and post-machined as a demonstrator for this study. For future design iterations, it is possible to further reduce the weight of the tracker by integrating hollow or lattice structures. In the next section, geometric distortions of the additively manufactured tracker were measured to assess the tolerances of the initial part.

Geometric Distortion Measurement

In a typical LPBF process, important parameters include and are not limited to laser power, laser speed, layer thickness, scan pattern, and build orientation. These process parameters have effects on the melting mode and the cooling rate of the

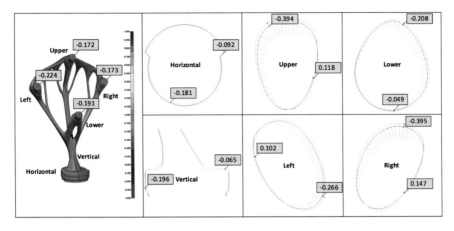

Fig. 11.12 Geometric distortion (mm) of the manufactured tracker (color) compared to the CAD design as a ground truth (grey)

material. Due to the repeated, rapid, and focused thermal cycle, residual stresses are accumulated in the printed parts, which then lead to geometric distortions and dimensional inaccuracies after part removal from the build plate [51]. In order to plan for machining allowance on the design, it is helpful to measure/predict the geometric distortion of the printed part in order to reduce the scrap.

Figure 11.12 shows that the scanned geometries aligned well with the CAD at the tracker base. A negative distortion value indicates a lack of material compared to CAD while a positive distortion value indicates an excess of material. A maximum shrinkage of about less than 0.2 mm in the radial direction was observed, as indicated by the geometric deviation measured along the horizontal and vertical cross sections on the base. The shrinkage may have resulted from the contraction of metal on cooling during the LPBF process. Based on these observations, through AM alone, the sample generated had a maximum deviation at the center of the posts of −224 μm with a maximum local dimensional distortion of 395 μm. Overall, the deviations at the tracker posts were smaller than 0.5 mm, which is an excellent performance outcome for such a tall structure in LPBF [52]. The good performance was, in part, the result of the interconnected posts at the top of the tracker, increasing the heat transfer and mechanical stability during the printing process. Based on these results, there is an opportunity to re-iterate in the design to compensate for such geometric inaccuracies, or to provide machining allowance to achieve the final tolerances. As a result, the 5 mm machining allowance added to the posts is considered to be sufficient to accommodate for machining to bring the final positions of the tips of the tracker posts back to the expressed tolerance criteria in Table 11.1.

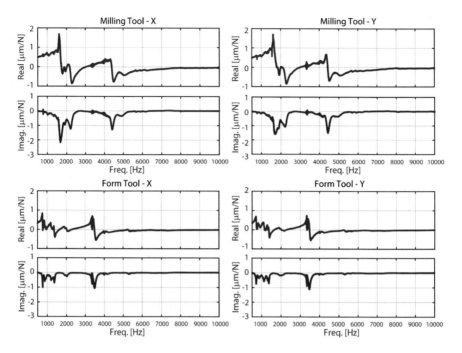

Fig. 11.13 FRF's of the milling tool and the form tool measured at the tool tip in the X- and Y-axis directions

Mechanical Vibrations

Milling Tool and Form Tool FRFs

The acceleration FRF's obtained for the 5-teeth Ø8mm milling tool and the 4-teeth Ø12mm form tool were converted to receptance as presented in Fig. 11.13. The largest negative peak of the real part, which is a critical region for machining stability, occurs at 2292 [Hz] and 2319 [Hz] for the milling tool in X- and Y-directions, respectively, while it occurs at 3575 [Hz] and 3543 [Hz] for the form tool in X- and Y-directions, respectively. Further implications of the tool FRFs on the machining stability are presented in a later section in terms of stability lobe diagrams and chatter frequencies.

Tracker FRFs

The magnitude of acceleration point FRFs measured from the tracker confined by different fixtures (stand, encapsulating ABS, adhesive) was converted to receptance and compared in Fig. 11.14. Overall, both methods of fixtures (ABS and adhesive) attenuated the vibration response compared to the stand configuration. However, at

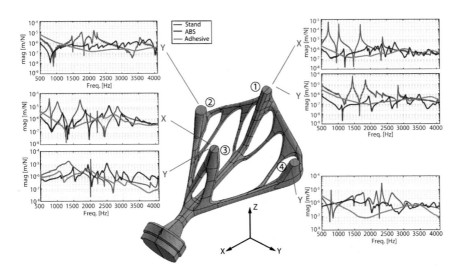

Fig. 11.14 Receptance FRFs of the tracker with different fixtures. Stand: Constrained only from the bottom with clamp, ABS: Sandwiched in the printed ABS negatives, Adhesive: Fixture using UV cured polymer

higher frequencies, the FRF magnitudes converge to similar values. Another main trend is the comparably flat shape of the FRF with the adhesive which can be attributed to the higher damping exerted by the adhesive material. These results have only partial implications on the machining stability as the latter depends on the combination of tool and workpiece FRFs, as well as the spindle speed under consideration. From the workpiece deflection viewpoint, which can result in an over-cut, the magnitude of the FRFs is important. In this case, the two clamping strategies seem to have similar properties, as their respective FRF magnitudes are either close or alternate in rank.

Machining Stability Diagrams

Machining stability diagrams obtained using frequency domain calculations in CutPRO® using the measured FRFs with the machining parameters mentioned in the earlier section on Frequency Domain Chatter Stability Analysis are shown in Figs. 11.15 and 11.16. The ultimate stability characteristic of each fixture (stand, ABS, adhesive) can be determined from the allowable depth of cut shown in the upper graph for each point. The chatter frequency is also presented. In terms of allowable depth of cut, it is observed that the encapsulating ABS performs the best for points 1, 2, 4, while the adhesive polymer fixture performs the best in the case of point 3. Although the adhesive has lower stiffness compared to the ABS contact, the modification of the fixture with a bolted bracket as shown in Fig. 11.5 at point 3 has imparted a large stiffness. For points 2 and 4, the adhesive fixture does some

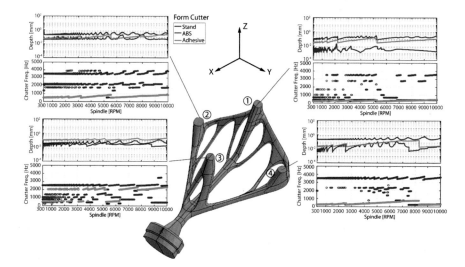

Fig. 11.15 Chatter stability diagram with the form tool

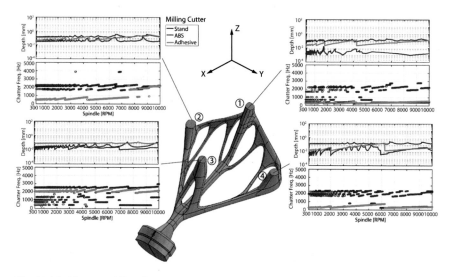

Fig. 11.16 Chatter stability diagram with the milling cutter

improvement overall when compared to the almost impossible to machine case of the stand configuration. As the workpiece FRFs are close, workpiece deflection during machining is expected to be similar. On the other hand, overall, ABS fixture can result in slightly better productivity as it is more stable in 3 of the 4 machined posts. Due to the closeness of the results, additional factors such as the availability of AM in the machine shop for ABS fixtures and the volume of the batch for UV polymer costs would determine the choice in practice.

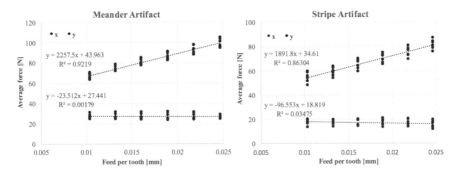

Fig. 11.17 Linear regression between average forces and feed per tooth

Table 11.7 Cutting coefficients identified for the stripe and meander artifacts

Strategy	K_{tc} (N/mm^2)	K_{rc} (N/mm^2)	K_{te} (N/mm)	K_{re} (N/mm)
Stripe	2779.9	1576.6	12.4	42
Meander	3234.2	2011.9	13	56.1

Cutting Tests and Machining

Identification of Cutting Force Coefficients

According to the average force linear regression method [53], average forces per tooth period and feed per tooth have a linear relationship. As a result, the measured average forces in the X and Y directions were plotted versus feed per tooth and then linear regression was applied (Fig. 11.17).

The slopes and y-intercepts of the fitted trend lines were identified, and they are functions of cutting force coefficients, cutting conditions, and cutter geometry. With known cutting conditions and cutter geometry, cutting force coefficients were solved and listed in Table 11.7. It can be observed that the meander artifact has higher cutting force coefficients than the stripe artifact. As a result, cutting forces were higher when machining the meander as shown in Fig. 11.17.

Tracker Inspection Results (ABS and UV)

The tracker, which was machined using the ABS enclosure and UV polymer adhesive fixture, was inspected on a coordinate measuring machine (CMM) using both tactile and fringe probes. The tracker was fixed on the CMM bed by hot gluing on 3 posts as shown in Fig. 11.18. The tactile probe was used for the initial alignment and measuring the positions/sizes of the posts and the magnet holes at the bottom. The fringe probe was used to generate the heat map images that trace the majority of the tracker surface. The metrology results are shown in Fig. 11.19 for the ABS enclosure (a) and UV adhesive polymer (b) cases. Deviation of individual circular features from the nominal for both cases is tabulated in Table 11.8.

Fig. 11.18 CMM measurement setup with the tracker hot glued on 3 posts on the CMM bed

Regarding all the measured features, including both the true position and the absolute value of the diameter deviation, the ABS enclosure seems to have larger errors ranging from +20% to +1067% compared to the UV polymer case. Only the true position of Datum Circle B constitutes an exception where the ABS enclosure has 82% less error. The qualitative distributions of the heatmaps shown in Fig. 11.19 also suggest more pronounced deviations in the ABS enclosure case. Overall, the ABS enclosure fixture seems to produce more dimensional inaccuracies which can be attributed to the larger clamping forces applied on the tracker. While this sandwiched configuration imparts greater stiffness to the workpiece, as discussed in the sections previously, it appears to distort the part geometry compared to the UV polymer case. This is further verified by the fact that the deviation in the UV polymer case is larger in the case of Datum Circle B located at post 3 (Fig. 11.14) where an additional clamping bracket was used to constrain the tracker. Hence, a tradeoff appears to exist in the fixture design between workpiece stiffness and part distortion due to clamping forces.

The complex structure designed and manufactured using the AM and SM workflow described in Fig. 11.2 resulted in a product demonstrator which was intended

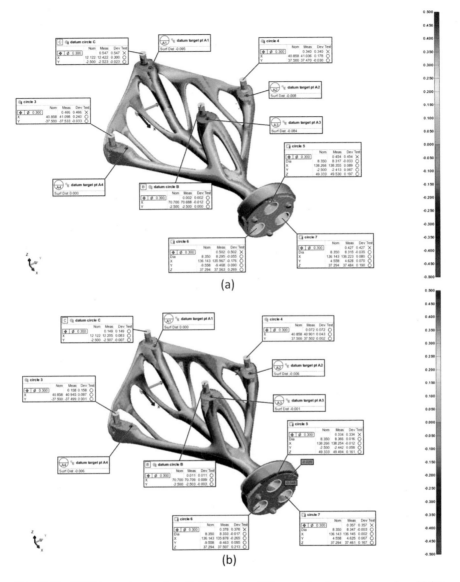

Fig. 11.19 CMM inspection results with the tactile and fringe probes for both cases of machining fixtures: (**a**) ABS enclosure, (**b**) UV polymer adhesive. Dimensions in mm

to adhere to the design criteria listed in Table 11.1. The design criteria 2 thru 8 were met. Through AM alone, the sample generated had a maximum local dimensional deviation of the tracker posts of 395 μm, which was not in compliance with the ±25 μm criterion for tolerance (criterion 1). The machining process was thus implemented using the two fixture types, one type being an AM enclosure, the other type being a UV polymer fixture. The ABS enclosure performed poorly, resulting in a

Table 11.8 CMM inspection results for ABS enclosure and UV polymer machining fixture cases. True position and absolute value of diameter deviation are presented for the inspected features

Feature	ABS enclosure (mm)	UV polymer (mm)	ABS change from UV polymer Δ%
Datum Circle B – true pos.	0.002	0.011	−82%
Datum Circle C – true pos.	0.547	0.149	+267%
Circle 3 – true pos.	0.466	0.158	+195%
Circle 4 – true pos.	0.340	0.072	+372%
Circle 5 – true pos.	0.454	0.334	+36%
Circle 6 – true pos.	0.502	0.378	+33%
Circle 7 – true pos.	0.427	0.357	+20%
Circle 5 – abs. Dia. dev.	0.033	0.016	+106%
Circle 6 – abs. Dia. dev.	0.055	0.017	+224%
Circle 7 – abs. Dia. dev.	0.035	0.003	+1067%

maximum local dimensional deviation at a single post of more than 500 μm due to deformation during clamping, thus not complying with criterion 1. The UV polymer fixture resulted in a maximum such deviation of 158 μm after machining, illustrating a significant improvement. The deviations reported in Circle 5, 6, and 7 in Fig. 11.19a, b, corresponding to the magnet clearance holes, were not part of the design criteria, as mounting the magnets does not require precision placement. In future work, it is recommended that the AM and SM workflow be re-iterated to explore increasing the volume fraction of the design space. This would increase the thickness of the struts connecting the four posts, while reducing the extra material allowance required for machining at the posts to reduce the product overall weight. Executing the AM distortion simulation to predict distortion and compensate for it in the design space iteratively until convergence would be needed. In the machining process, different types of UV polymers could be explored to improve the stiffness and vibration response and obtain a better machining tolerance.

Conclusions and Reflections on Additive Manufacturing and Post-process Machining

In this chapter, a workflow which synergizes the AM and post-machining processes was proposed and deployed in a case study of manufacturing a surgical tool using LPBF followed by 5-axis milling. Using AM, the original tracker assembly was consolidated into one piece such that the high part rejection rate due to the accumulation of dimensional inaccuracy and resulting tolerance violation during the assembly process were prevented. By post-machining the mating surfaces of the tracker posts, the surface roughness was reduced and dimensional inaccuracies due to geometric distortion in AM were corrected to fulfil functional requirements. Key findings regarding the synergistic process are as follows:

1. The design space, load cases, minimal filter radius, and manufacturing constraints can all affect the results of structural topology optimization. The most suitable design should be selected among all candidates considering functional and manufacturing requirements.
2. Unit cost, thermally induced distortion, and support structures were reduced through build orientation optimization.
3. Workholding of complex AM parts is challenging. The encapsulating ABS fixture provided good stiffness and damping of the tracker during machining, but the flexible tracker was pre-deformed under the clamping force. The UV system addressed the pre-deformation issue; however, the accessibility of the UV device to some regions of the part was limited.
4. The final maximum dimensional deviation at the posts was 395 μm via AM, with the ABS encapsulation resulting in a maximum deviation of 547 μm, while the UV polymer-based fixture had a better performance, with 158 μm respectively. The final geometric fidelity of the product can be further improved by iterating through the Workflow.

The fabrication of complex-shaped components with a high geometric tolerance requires a complementary knowledge of both additive manufacturing (AM) and subtractive machining (SM) processes to optimize the design and manufacturing synergistically. In the future, these options can be incorporated in the synergistic process for light-weighting. The experience gained through this collaborative project has opened an appreciation for the level of dedication, expertise, and care required to go from powder to product. More importantly, it was an excellent opportunity to learn from a cross disciplinary effort.

Acknowledgments The authors appreciate the funding support received from The Natural Sciences and Engineering Research Council of Canada (NSERC) – Canadian Network for Research and Innovation in Machining Technology (CANRIMT2), grant number NETGP 479639-15 and the funding contribution provided by the FedDev Ontario (Program #809104). This work was accomplished in partnership with Renishaw (Canada) Solutions Centre (Kitchener, ON, Canada), Intellijoint Surgical (Waterloo, ON, Canada), and Blue Photon (Shelby, MI, USA). In addition, the authors would like to acknowledge the help of Jerry Ratthapakdee and Karl Rautenberg at the University of Waterloo and Carl Hamann at Renishaw in helping with the LPBF prints, Robert Wagner at the University of Waterloo in helping with sample machining and cutting force measurements, and the motivation and support of the Multi-Scale Additive Manufacturing (MSAM) Group at the University of Waterloo.

References

1. Liu, J.: Guidelines for AM part consolidation. Virtual Phys. Prototyp. **11**(2), 133–141 (2016)
2. Schmelzle, J., Kline, E.V., Dickman, C.J., Reutzel, E.W., Jones, G., Simpson, T.W.: (Re)designing for part consolidation: understanding the challenges of metal additive manufacturing. J. Mech. Des. **137**(11), 111404 (2015)

3. Yang, S., Tang, Y., Zhao, Y.F.: A new part consolidation method to embrace the design freedom of additive manufacturing. J. Manuf. Process. **20**, 444–449 (2015)

4. Rosen, D.W.: Computer-aided design for additive manufacturing of cellular structures. Comput.-Aided Des. Appl. **4**(5), 585–594 (2007)

5. Murr, L.E., et al.: Metal fabrication by additive manufacturing using laser and electron beam melting technologies. J. Mater. Sci. Technol. **28**(1), 1–14 (2012)

6. Murr, L.E., et al.: Next-generation biomedical implants using additive manufacturing of complex, cellular and functional mesh arrays. Philos. Trans. R. Soc. Math. Phys. Eng. Sci. **368**(1917), 1999–2032 (2010)

7. Zegard, T., Paulino, G.H.: Bridging topology optimization and additive manufacturing. Struct. Multidiscip. Optim. **53**(1), 175–192 (Jan. 2016)

8. Brackett, D., Ashcroft, I., Hague, R.: Topology optimization for additive manufacturing. In: Proceedings of Solid Freeform Fabrication Symposium, pp. 348, 15–362, Austin, TX (2011)

9. Langelaar, M.: Topology optimization of 3D self-supporting structures for additive manufacturing. Addit. Manuf. **12**, 60–70 (2016)

10. Gaynor, A.T., Guest, J.K.: Topology optimization considering overhang constraints: eliminating sacrificial support material in additive manufacturing through design. Struct. Multidiscip. Optim. **54**(5), 1157–1172 (2016)

11. Langelaar, M.: An additive manufacturing filter for topology optimization of print-ready designs. Struct. Multidiscip. Optim. **55**(3), 871–883 (2017)

12. Su, X., Yang, Y., Yu, P., Sun, J.: Development of porous medical implant scaffolds via laser additive manufacturing. Trans. Nonferrous Met. Soc. China. **22**, s181–s187 (2012)

13. Tuck, C.J., Hague, R.J.M., Ruffo, M., Ransley, M., Adams, P.: Rapid manufacturing facilitated customization. Int. J. Comput. Integr. Manuf. **21**(3), 245–258 (2008)

14. Pallari, J.H.P., Dalgarno, K.W., Munguia, J., Muraru, L., Peeraer, L., Telfer, S.: Design and additive fabrication of foot and ankle-foot orthoses. In: Proceedings of the 21st Annual International Solid Freeform Fabrication Conference – An Additive Manufacturing Conference, p. 12, Austin, TX, USA (2010)

15. Tuomi, J., et al.: A novel classification and online platform for planning and documentation of medical applications of additive manufacturing. Surg. Innov. **21**(6), 553–559 (2014)

16. Salmi, M., et al.: Patient-specific reconstruction with 3D modeling and DMLS additive manufacturing. Rapid Prototyp. J. **18**(3), 209–214 (2012)

17. Wang, X., et al.: Topological design and additive manufacturing of porous metals for bone scaffolds and orthopaedic implants: a review. Biomaterials. **83**, 127–141 (2016)

18. Linxi, Z., Quanzhan, Y., Guirong, Z., Fangxin, Z., Gang, S., Bo, Y.: Additive manufacturing technologies of porous metal implants. China Foundry. **11**, 322–331 (2014)

19. 3D Printed Surgical Tool Inspired By Origami. All3DP, 06-Apr-2016. [Online]. Available: https://all3dp.com/3d-printed-surgical-tool-inspired-origami/. Accessed 20 Mar 2019

20. New 3D Printed Medical Tool a Breakthrough for ACL Reconstruction Surgery. 3D Printing Industry, 08-Jan-2016. [Online]. Available: https://3dprintingindustry.com/news/new-3d-printed-surgical-tool-a-breakthrough-for-acl-reconstruction-surgery-64519/. Accessed 20 Mar 2019

21. Bernhard, J.-C., et al.: Personalized 3D printed model of kidney and tumor anatomy: a useful tool for patient education. World J. Urol. **34**(3), 337–345 (2016)

22. AlAli, A.B., Griffin, M.F., Butler, P.E.: Three-dimensional printing surgical applications. Eplasty. **15** (2015)

23. Manogharan, G., Wysk, R., Harrysson, O., Aman, R.: AIMS – a metal additive-hybrid manufacturing system: system architecture and attributes. Procedia Manuf. **1**, 273–286 (2015)

24. Davim, J.P. (ed.): Surface integrity in machining. Springer, New York/London (2009)

25. Fox, J.C., Moylan, S.P., Lane, B.M.: Effect of process parameters on the surface roughness of overhanging structures in laser powder bed fusion additive manufacturing. Procedia CIRP. **45**, 131–134 (2016)

26. Mumtaz, K., Hopkinson, N.: Top surface and side roughness of Inconel 625 parts processed using selective laser melting. Rapid Prototyp. J. **15**(2), 96–103 (2009)
27. Jamshidinia, M., Kovacevic, R.: The influence of heat accumulation on the surface roughness in powder-bed additive manufacturing. Surf. Topogr. Metrol. Prop. **3**(1), 014003 (2015)
28. Gong, H., Rafi, K., Gu, H., Starr, T., Stucker, B.: Analysis of defect generation in Ti–6Al–4V parts made using powder bed fusion additive manufacturing processes. Addit. Manuf. **1–4**, 87–98 (2014)
29. Mower, T.M., Long, M.J.: Mechanical behavior of additive manufactured, powder-bed laser-fused materials. Mater. Sci. Eng. A. **651**, 198–213 (2016)
30. Gong, H., Rafi, K., Gu, H., Janaki Ram, G.D., Starr, T., Stucker, B.: Influence of defects on mechanical properties of Ti–6Al–4V components produced by selective laser melting and electron beam melting. Mater. Des. **86**, 545–554 (2015)
31. Gong, H., Rafi, K., Starr, T., Stucker, B.: Effect of defects on fatigue tests of as-built Ti-6AL-4V parts fabricated by selective laser melting. In: Proceedings of the Solid Freeform Fabrication Symposium, pp. 499–506, Austin, TX, USA (2012)
32. Frazier, W.E.: Metal additive manufacturing: a review. J. Mater. Eng. Perform. **23**(6), 1917–1928 (2014)
33. Niendorf, T., Leuders, S., Riemer, A., Richard, H.A., Tröster, T., Schwarze, D.: Highly anisotropic steel processed by selective laser melting. Metall. Mater. Trans. B Process Metall. Mater. Process. Sci. **44**(4), 794–796 (2013)
34. Paul, R., Anand, S., Gerner, F.: Effect of thermal deformation on part errors in metal powder based additive manufacturing processes. J. Manuf. Sci. Eng. **136**(3), 031009 (2014)
35. Li, C., Liu, J.F., Guo, Y.B.: Efficient multiscale prediction of cantilever distortion by selective laser melting. In: Proceedings of the 27th Annual International Solid Freeform Fabrication Symposium-An Additive Manufacturing Conference, pp. 236–246
36. Patil, N., et al.: A generalized feed forward dynamic adaptive mesh refinement and derefinement finite element framework for metal laser sintering—part I: formulation and algorithm development. J. Manuf. Sci. Eng. **137**(4), 041001 (2015)
37. Wang, D., Yang, Y., Yi, Z., Su, X.: Research on the fabricating quality optimization of the overhanging surface in SLM process. Int. J. Adv. Manuf. Technol. **65**(9–12), 1471–1484 (2013)
38. R. plc, Renishaw: Design for metal AM – a beginner's guide. Renishaw. [Online]. Available: http://www.renishaw.com/en/design-for-metal-am-a-beginners-guide%2D%2D42652. Accessed 13 Feb 2019
39. Thomas, D.: The Development if Design Rules for Selective Laser Melting. Univ. Wales (2009)
40. Vora, P., Mumtaz, K., Todd, I., Hopkinson, N.: AlSi12 in-situ alloy formation and residual stress reduction using anchorless selective laser melting. Addit. Manuf. **7**, 12–19 (2015)
41. Xu, W., et al.: Additive manufacturing of strong and ductile Ti–6Al–4V by selective laser melting via in situ martensite decomposition. Acta Mater. **85**, 74–84 (2015)
42. Zhao, X., Lin, X., Chen, J., Xue, L., Huang, W.: The effect of hot isostatic pressing on crack healing, microstructure, mechanical properties of Rene88DT superalloy prepared by laser solid forming. Mater. Sci. Eng. A. **504**(1–2), 129–134 (2009)
43. Tammas-Williams, S., Withers, P.J., Todd, I., Prangnell, P.B.: The effectiveness of hot isostatic pressing for closing porosity in titanium parts manufactured by selective electron beam melting. Metall. Mater. Trans. A. **47**(5), 1939–1946 (2016)
44. Sodick, OPM250L – Metal 3D Printer. Sodick, 10-Feb-2019. [Online]. Available: //www.sodick.com/products/metal-3d-printing/opm250l. Accessed 11 Feb 2019
45. LASERTEC 65 3D hybrid – ADDITIVE MANUFACTURING Machines by DMG MORI." [Online]. Available: https://en.dmgmori.com/products/machines/additive-manufacturing/powder-nozzle/lasertec-65-3d-hybrid. Accessed 11 Feb 2019
46. Boivie, K., Karlsen, R., Ystgaard, P.: The concept of hybrid manufacturing for high performance parts#. South Afr. J. Ind. Eng. **23**(2) (2011)

47. Meeting the Machining Challenges of Additive Manufacturing. [Online]. Available: https://www.mmsonline.com/articles/meeting-the-machining-challenges-of-additive-manufacturing. Accessed 03 Apr 2019
48. Oyelola, O., Crawforth, P., M'Saoubi, R., Clare, A.T.: Machining of additively manufactured parts: implications for surface integrity. Procedia CIRP. **45**, 119–122 (2016)
49. "DirectAdditiveSubtractiveHybridManufacturing.pdf."
50. Budak, E., Altintaş, Y.: Analytical prediction of chatter stability in milling—part I: general formulation. J. Dyn. Syst. Meas. Control. **120**(1), 22 (1998)
51. Cheng, L., A. To: Part-scale build orientation optimization for minimizing residual stress and support volume for metal additive manufacturing: theory and experimental validation. Comput. Aided Des. **113**, 1–23 (Aug. 2019)
52. Li, C., Liu, Z.Y., Fang, X.Y., Guo, Y.B.: Residual stress in metal additive manufacturing. Procedia CIRP. **71**, 348–353 (2018)
53. Altintas, Y.: Manufacturing automation: metal cutting mechanics, machine tool vibrations, and CNC design, 2nd edn. Cambridge University Press, New York (2012)

Yanli Zhu is a talented engineer, with a passion for control and optimization of manufacturing processes, with a focus on synergies between machining and additive manufacturing processes. She is a recent graduate with a master of applied science in mechanical and mechatronics engineering program at the University of Waterloo. The research highlighted in this present work stems from her thesis work, where she focused on developing a process workflow complementing additive manufacturing (AM) and post-machining in order to fabricate complex geometries. Her passionate skills in design for additive manufacturing, topology optimization, process parameter optimization, simulation of thermomechanical processes, and machining processes, as well as hands-on experimental work in both machining and additive manufacturing, have brought to life a complex biomedical tool presented in the chapter.

Ahmet Okyay is presently a mechanical design engineer at ASML, the Netherlands. The research highlighted in this present work stems from his efforts toward the characterization of vibration response during machining of complex-shaped additively manufactured components, in his role as a research associate in the Precision Control Laboratory at the University of Waterloo. His main research interests span the mechatronic design of actuators for precision positioning and dynamic damping applications, with an emphasis on the mechanical design, electromagnetics, controls, modal analysis, and error budgets.

Mihaela Vlasea started on her journey in the additive manufacturing world in 2008 by embarking on her PhD degree in mechanical and mechatronics engineering at the University of Waterloo in the area of additive manufacturing (AM). During her PhD, she built a hybrid binder jetting machine for the production of metal and ceramic components, with an initial goal to fabricate bone augmentation and replacement components. In 2014, she finally got to see the machine in action, where in collaboration with the University of Toronto, Mount Sinai Hospital, and the University of Waterloo she designed and manufactured custom ceramic implants for in vivo trials. In 2015, Mihaela continued her work on building AM mechatronic systems through the design of an open platform laser powder bed fusion (LPBF) test bed at the National Institute of Standards and Technology (NIST), Gaithersburg, US. Mihaela joined the Mechanical and Mechatronics Engineering Department (MME) at the University of Waterloo as an assistant professor in Sept 2015 and is the research co-director of the Multi-Scale Additive Manufacturing Lab. She gets to work with some of the most talented students, researchers, and staff on bridging the technological gaps necessary to improve AM part quality, process repeatability, and reliability for powder bed fusion and binder jetting AM processes. One of the most rewarding aspects of her work is to translate science into reality by engaging with numerous industry partners across sectors. Her quality of work is patiently supervised by two cats and two dogs, who have read over her shoulder some of the most exciting research theses out there. But most importantly, her career path has been shaped by her family and by the wonderful additive manufacturing community.

Kaan Erkorkmaz is a professor in the Department of Mechanical and Mechatronics Engineering at the University of Waterloo. He is a fellow of the International Academy for Production Engineering (CIRP) and also a registered professional engineer in the Province of Ontario. Professor Erkorkmaz's research interests are in the areas of precision manufacturing, machining, and dynamics; controls, and optimal trajectory planning for machine tools; with the objective of increasing the productivity, part quality, and resource efficiency in manufacturing operations.

Mark Kirby is presently the industry training manager at the Multi-Scale Additive Manufacturing Lab, University of Waterloo. The research highlighted in this present work stems from his efforts toward industrialization and adoption of additive manufacturing processes in his role as the additive manufacturing business manager at Renishaw and business development manager at ADEISS. Mark is an experienced manufacturing manager with a demonstrated history of working in the aerospace and additive manufacturing industry. He has a great deal of hands-on manufacturing experience, analytical skills, Computer-Aided Design (CAD), CAM, and quality management (AS9100 and ISO13485) and is overall a passionate champion of sensible additive manufacturing tech adoption.

Chapter 12
Additive as an Entrepreneur and Outreach Tool

Erin Winick Anthony

Starting a 3D Printing Company in College as an Engineering Undergrad

"You'll learn more starting a company yourself than getting a minor in entrepreneurship."

These words of advice given to me by one of my college professors are the ones that shaped my final years as an undergraduate mechanical engineering major at the University of Florida.

I chose mechanical engineering as my major because, like many engineers, I was passionate about designing and making things. As a kid I built massive LEGO towers (Fig. 12.1) and marble runs in my parents' living room, loved sewing and crafts, and folded an origami dragon out of any spare pieces of paper I could find lying around.

But during my college career, I found my passion deviating from that of my peers. Those around me were working towards research gigs that would get them into graduate programs, and industry connections that would get them a full time offer at a Fortune 500 company after graduation. While my four industry internships were great learning experiences, that wasn't where I saw my future headed.

3D printing was my tool to forge my own path ahead. It was my key to starting my own company.

E. W. Anthony (✉)
Sci Chic, Gainesville, FL, USA
e-mail: erin.winick@gmail.com

© Springer Nature Switzerland AG 2021
S. M. DelVecchio (ed.), *Women in 3D Printing*, Women in Engineering and Science, https://doi.org/10.1007/978-3-030-70736-1_12

Fig. 12.1 Erin playing with LEGO and dressup in her childhood home in Florida

The First Layer

I was first exposed to 3D printing in my freshman year design class in 2012. It immediately caught my interest. Seeing the gears that I had designed on the computer turn into gears I could spin in my hand was magical.

Before this manufacturing had felt distant and inaccessible. Something you watched on TV shows about how motors and engines were made. But 3D printing was something I could do myself. I didn't need a factory floor at my disposal. I began seeking out opportunities to learn more, and get more involved in the manufacturing field as a whole. It served as the gateway to using those manufacturing floor tools such as lathes, mills, and bandsaws, I had seen on those TV shows.

I got my first chance to dive into this new interest during my summer engineering internships. I saw how large companies were adapting to the rise of additive manufacturing, the improvement of the tools, and the reduction of price in 3D printers.

Although 3D printing has been around for decades, I had the luck of going through college as the technology was becoming more prevalent, and more companies were purchasing new printers.

My sophomore year I interned at Solar Turbines in San Diego, California, in the tool design area. I got an increased appreciation for the capabilities of 3D printing beyond just what an $800 printer I bought off of Amazon could accomplish. I saw metal printing in action, and learned about new soft materials that were coming to the market.

The real point of inspiration for my company came during Introduce a Girl to Engineering Day and SWE 3D Printing day. These events were hosted by the Society of Women Engineers University of Florida Section the year I served as the section's president. During the events, I witnessed children experiencing the same awe that I had with the technology (Fig. 12.2). A nametag they were creating in minutes on the computer was being created in colorful plastic. Not only were they being exposed to manufacturing at a young age, but also the bridges between engineering, creativity, and art.

This got my gears turning. There had to be a way to scale this experience. And by doing so, might I be able to find my career niche that I had struggled with finding? I started transitioning the manufacturing and 3D printing lessons I was learning in my classes and industry internships into my own project, a project that could help others experience the wonders of 3D printing in the same way I had.

That is how Sci Chic was born.

Filament in Fashion

Sci Chic uses plastic and metal 3D printing to create science- and engineering-inspired jewelry for kids and adults. The goals were to use the jewelry to serve as a tangible way to expose kids to 3D printing, demonstrate that science can be beautiful and creative, and to create stylish pieces of jewelry that would spark conversations about science. I was combining my Project Runway, Shark Tank, and manufacturing floor dreams all into one.

Combining 3D printing and fashion might not seem like an immediately obvious choice, but the fashion industry has been embracing it more in recent years. The Met Gala is an annual event which features one of the most innovative, glamorous, and wild nights of fashion. The 2018 event showed off five dresses and headpieces at least partially made from 3D printing [1]. They even included a dress made of large custom printed flower petals designed by Zac Posen that took more than a thousand hours to print and assemble. All of the women wearing the 3D printed dresses were scanned to ensure the creation fit them perfectly.

Posen is known for embracing technology in his fashion. He also created the gorgeous work of art that is the now famous dress Claire Danes wore to the 2016 Met Gala. It was intricately created from fiber optic fabric, forming a dress that would have been worn by Cinderella if she had minored in engineering.

Fig. 12.2 One of the participants at the University of Florida Society of Women Engineers 3D printing outreach events

Israeli fashion designer Danit Peleg also helped pioneer 3D printed fashions, creating the world's first collection to be entirely created using home desktop 3D printers in 2015 [2]. She also launched the first customizable 3D printed garment available for online purchase in 2017, a jacket available for a hefty $1500. Her clothes benefit from being designed to be more flexible than the rigid sculptures of runways past.

These dresses and clothing items exemplify how 3D printing and fashion best mesh: the creation of intricate, custom, artistic pieces. They capture the imagination of the public, drawing attention to the art of additive manufacturing. The technology can enable new forms, designs, and styles for the runway previously not possible. But those designs don't come cheap or easy to wear. Zac Posen's flower petal dress weighed about 30 pounds. That's definitely not every day office wear.

Sci Chic was born in the midst of this innovation in the fashion industry, with a desire to make smaller pieces of 3D printed style that could be worn on a daily basis. I didn't want our price points to be in the thousand dollar and up range. That's not an avenue to reach a large audience and make the biggest impact in education. So I focused on designs that could be made more quickly, and included less alterations to each item.

Designing a Company

Sci Chic was not born overnight. The idea came even before I purchased my own 3D printer. First I had the desire to combine my skills and areas of interest.

I grew up a fan of sewing, creating Halloween costumes and outfits with my mom behind her sewing machine. A sewing machine was the first true manufacturing machine I ever got experience using, and it helped me learn to turn something 2D (fabric) into a 3D creation I could wear.

At the same time, I spent all of my time building model roller coasters, and watching science entertainment like Bill Nye and Mythbusters.

Sci Chic was my outlet to use manufacturing and fashion as a science communication tool.

The first work I did on the company was modeling the designs. I hopped onto my engineering design software I had learned in my classes and created some of the initial core patterns that I would use for years to come: the Atom Necklace, Moon Phase Necklace, and Trajectory Necklace as seen in Fig. 12.3.

The Atom Necklace was inspired by Bohr's Model of an atom, the Moon Phase Necklace showed the phases of the moon evolving vertically, and the Trajectory Necklace depicted the path from Earth to Moon that the Apollo 11 astronauts followed, also known as translunar injection.

Fig. 12.3 Three of Sci Chic's first jewelry designs. From left to right: White plastic Atom Necklace, stainless steel Moon Phase Necklace, and the stainless steel Apollo Trajectory Necklace

Fig. 12.4 Erin working in her apartment to create blue plastic jewelry for sale on the Sci Chic website. She is wearing a Wheatstone Bridge Necklace and Earrings

These concepts went from ideas to designs to for-sale in an online shop very quickly. And 3D printing was what allowed that to happen. The creation of a minimum viable product (MVP) is crucial to most successful startups. If you can sell that MVP, it's even better.

The investment in the purchase of my first Lulzbot Mini 3D printer made the creation of a MVP extremely easy. My college apartment transformed into a factory (Fig. 12.4). I made designs, printed them out, and assembled necklace chains or earrings hooks onto the print. Then I gave them a test run.

If the hole in the pendant was too small for the chain, or the necklace wasn't balanced, I could immediately create tweaks to the design, and test it out again. It even let me try out new ideas like 3D printing my own designs onto fabric. While 3D printed fabric didn't turn into a product I could sell on my website, I was able to create a skirt for wearing to events and showing off the technology. It was more akin to a fashion runway piece than something I could mass produce.

Additive manufacturing made the design part of Sci Chic simple. I didn't have to loop in a supplier to test every single item. This allowed us to start selling items without waiting for shipment to come in.

Scaling the Model

Prototyping was done. The first designs were made. It was time for launch. Everything kicked off at the Harn Museum Art in Engineering event in 2015, the same day our online store was launched. Clad in a galaxy print dress with printers whirring behind me, I got to see the first reaction to this idea, and dip my toe into the world of entrepreneurship.

Then came the hard work that many startup founders go through. Keeping the momentum going and growing the company while actually producing the product. There were long nights of working on the website, running the 3D printers, assembling jewelry, printing shipping labels, responding to customer service emails, and more. All of that was great at the start, but we had to scale.

3D printing excels at customization and control, but speed and mass production isn't its forte. Especially when you are only using small desktop printers. 3D printing is constantly improving when it comes to speed, quality, and cost of machines, but other methods of manufacturing still beat it in a number of areas. In my case, the use of additive manufacturing was an education tool I did not want to pivot away from, but entrepreneurs should be willing to consider other manufacturing techniques like injection molding and traditional manufacturing techniques if they begin to enter a mass production phase.

I was able to strike a good balance by pulling into my supply chain a 3D printing farm to help me produce larger numbers of plastic pendants, and 3D printing company Shapeways for metal production.

3D printing farms are large numbers of additive manufacturing machines all run by one group. They offer a larger volume of production than companies that own only a few large machines, or makers that have a desktop printer. By pulling in Shapeways, I was able to add product variety. Shapeways offers materials like stainless steel, and services like metal polishing, creating pieces of jewelry more akin to the style adults would expect. Shapeways also provided software for checking to ensure my designs were 3D printable in the materials I was requesting.

These organizations gave me the resources to fill large orders, offer more variety, and, in some cases, offer higher quality than I could produce myself. My personal printers provided relative speed, while these outside companies provided quantity (Fig. 12.5).

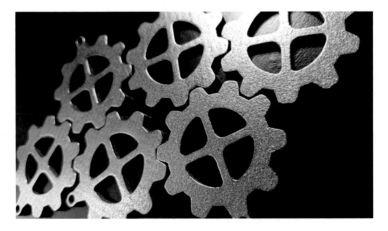

Fig. 12.5 Sci Chic gear pendants 3D printed in polished stainless steel through Shapeways

The addition of the 3D printing farms was especially important as we began producing hundreds of pieces of jewelry a month. In 2016, we were reaching more than $5000 in monthly sales, we had a subscription box running, and on top of that, I was graduating from college.

Other areas of support came from the resources available from living in a college town. While starting a company when in school can be daunting, it's also an amazing opportunity. Colleges want to support the entrepreneurial spirit in their community, and in turn, around your company. Many of the colleges will help promote your business in alumni magazines and on social media, provide support for entering business plan and pitch competitions, and allow easy access to 3D printers for prototyping. At the University of Florida, I developed great relationships with our science library which housed a number of 3D printers for public use. While I didn't use them for production of jewelry, they were helpful for testing out my ideas before purchasing a printer and for learning the technology.

I also entered a number of competitions in the local community and at other schools. And some of my travel was supported by the university. I won the elevator pitch competition at Texas Christian University's Values and Ventures Competition, a local SCORE award, and the SheKnows Campus Pitch competition. Not only did these provide some monetary support for the business, but they also helped me make connections with more people in similar fields that wanted to support my work. I highly recommend college entrepreneurs take advantage of opportunities like this, not only to potentially earn money to support your company, but also to force yourself to hone your pitch and message. There's nothing like getting up in front of a crowd and explaining your idea, and defending it against questions from judges (Fig. 12.6).

Being in a college town also meant the availability of many startup incubators that could support the business. Incubators can offer a number of things, depending on their focus. Typically they include things like office space, training, sales support, and even, once again, access to manufacturing and prototyping equipment. Although I didn't get the chance to take advantage of it, there was even a dorm at the University of Florida that housed a space for student use with laser cutters, 3D printers, and a number of other technologies. Students in the dorm were encouraged to use their rooms as office space to kick off a company.

For me, the better option came with moving into the Selling Factory, an incubator in Gainesville, Florida designed around scaling businesses, and helping those businesses sell their product. The biggest advantage to heading to an incubator was finally moving my 3D printers out of my apartment (Fig. 12.7). Separating my manufacturing and work space from my home was crucial to be able to separate my work and personal time. The apartment did seem strangely quiet though when the sound of 3D printers was no longer constantly running in the background.

From the experiences at the peak of our business, I compiled the main takeaways I gained for entrepreneurs considering using additive manufacturing:

- 3D printing is fantastic for testing, designing, and prototyping your idea.
- People respond to and get excited by 3D printing. It is a great way to hook the public and get them excited by your product.

Fig. 12.6 Erin at Texas Christian University after winning the elevator pitch competition at the Values and Ventures business plan competition

- 3D printing allows for a high level of control and customization. Low volume, custom products are often a good match for the process.
- When you are attempting to sell your product in large quantities, it sometimes makes sense to pivot to another manufacturing technique during your company's evolution.
- In cities, there are often local startup incubators, universities, or libraries that have 3D printers you can use. Take advantage of the resources in your community to create prototypes, learn about the technology, and see if 3D printing is a good fit.

Fig. 12.7 Erin working in the Selling Factory in Gainesville, Florida

Printing an Impact

At the same time I was learning these entrepreneurship lessons, I was also gaining knowledge about STEM outreach, the experiences that would fuel my career.

I hauled my original 3D printer around the United Stations in the backseat of my car and in the cargo storage of buses. My 3D printers are probably some of the most well-traveled ones out there, making numerous cross country trips. I represented Sci

Chic and manufacturing at in person events from the USA Science and Engineering Festival in Washington DC to the Orlando Maker Faire.

3D printing is an effective outreach tool because of its portability and accessibility. Event participants can clearly see the material being used and the process being used to create the final product. Outreach events let you see 3D printing through fresh eyes. You get to experience the magic of this technology through the questions of a toddler, the conversation of a middle schooler who has used their library's printer, and parents of a child who loves art and wants to explore printing.

Being immersed in the world of additive manufacturing, it's easy to forget that most people still haven't seen a 3D printer in person. Some people still haven't even heard of one. By making the effort to attend these events with a 3D printer on my table and answering attendees' questions, I was really making an impact. I was exposing new audiences to manufacturing and the intricacies of 3D printing, and possibly inspiring a new generation of engineers and makers.

If you want to use a 3D printer at your outreach events, here are my tips:

1. *Use an enclosed printer, or keep it in a safe location:* 3D printers have very hot parts and hot plastic, and can be tempting for kids to touch. Using a printer with clear plastic surrounding the dangerous parts makes an accident much less likely. If that isn't an option, be sure small hands can't reach the hot parts of the machine.
2. *Make sure your table setup has access to electricity:* This may seem obvious, but a lot of events try to conserve tables with access to power outlets, and you may need to make a special request. 3D printers definitely aren't as cool when they aren't running.
3. *Before attending the event, be sure to print out at least one of the models that can sit on the table next to the printer:* You will get a lot of visitors at your booth interested in seeing the print when it finishes. This preprinted model also helps them see multiple parts of the process and lets them have something that can pick up and examine. If possible, having a print they can take with them as a giveaway is even better. Touching prints allows for conversations about topics like layer height in FDM printing, and the detail of prints.
4. *Pick a 3D model that is interesting and quick to print:* Despite having a finished print on the table, many people will want to come back to see the print complete. An 8-hour print is a big investment, and not many people are likely to remember to return before they leave. For me, I would always select a small piece of jewelry to print that would take around 30 to 60 minutes.
5. *Have someone available to engage with people as they watch the printer:* The printer will serve as the draw to the table, but make sure to capitalize on that opportunity by talking with the visitors to give them more information. Great conversation starters include:

    (a) Have you seen or used a 3D printer before?
    (b) Do you have a guess of what this machine is making?
    (c) Did you know 3D printers could be used to create X? Add in whatever you currently have on the printer.

6. *Bring extra filament, resin, or base material with you:* This is good to have on hand in case you print a lot of items, but it's also a good object for guests to interact with. They can see the material that feeds into the printer, and you can then explain how the printer melts or forms it into its final shape.

Print Complete

After about five years, I dialed Sci Chic back, but the lessons I learned from the experience have been invaluable. Starting and running this company are the reasons I am where I am today. It showed me my love of communicating science with the public, and building projects from the ground up. I am glad I followed my professor's advice back in college to pursue the creation of this company. I continue to use the lessons I have learned about business and getting the public excited about science and manufacturing. I now work full time at NASA's Johnson Space Center in Houston, Texas, as a science communications specialist for the International Space Station. I tell the public the amazing stories of the science being conducted in microgravity aboard the space station, and how it's improving life here on Earth.

I still have my 3D printers in my house and fill a jewelry order here or there, but my focus is more on personal projects and sharing them with the world through the internet. Just as I attempted to do with Sci Chic, the social media and online communities offer a way to scale the impact of my design and 3D printing work. I can reach the people that might never be able to go to those in person events.

For Sci Chic, I primarily used YouTube to extend our outreach. I created timelapses and behind the scenes videos while creating our jewelry, sort of like mini *How It's Made* episodes. This way my customers could see how 3D printing was used to make their necklace or earrings.

The main platforms I use now are Instagram and TikTok, with a dash of Twitter and Facebook thrown in (@erinwinick). Here I share posts featuring the computer-aided design process, 3D printing, and the final product. TikTok is great for short timelapses and clips from different parts of the making process. With voiceovers on the videos, I can give context and more information about manufacturing. Instagram excels when it comes to high-quality photos of the printing process and the final product, as well as livestreams of the printing process.

Through pictures and videos, I have shared the creation of jewelry, pieces of my Ms. Frizzle costume (Fig. 12.8), cell-inspired lamps, a 3D printed fabric skirt, a badge from Star Trek, and more. Outside of Sci Chic, my most famous project, and the one I am most proud of, is the sharing of how I 3D printed my wedding.

My husband and I wanted to make our wedding about who we are as people, and as a couple. What better tool than 3D printing to customize our wedding day. I set off creating designs from scratch, altering existing designs, and using models I found online. Over the course of more than 100 hours [3], I 3D printed and assembled my wedding bouquet, the bridesmaids wedding bouquets, our table numbers, the cake toppers and decorations, and my head band, all using desktop printers

Fig. 12.8 Erin wearing the Saturn earrings she 3D printed as a part of her Ms. Frizzle costume

Fig. 12.9 Pieces of Erin's 3D printed wedding (from left to right) – flowers for the bouquets, on-screen design of table numbers, 3D printed the table numbers, and Erin with her bouquet and 3D printed headband

(Fig. 12.9). Throughout all of that, I took my social media followers along for the ride. They saw behind the scenes clips, the bouquet assembly, the design process, and more.

By the time the big day arrived, everyone was just as (or even more) excited to see the final 3D printed products, as they were to see us actually get married. I was then able to reach an even bigger audience by writing about my wedding for MIT Technology Review [3]. The article got picked up by a number of other outlets who wanted to share this project as well.

I share this story to show that using your personal connection to the things you make and your projects can inspire others, and teach them about something new. There is a strong maker community online, and they create more incredible projects that spark the interest of groups outside of the typical audience that regularly consumes science and engineering information.

But the first step to encouraging others to create is putting yourself and your 3D printed projects out into the world. From writing a LinkedIn post depicting metal printers on the manufacturing floor creating aerospace parts to putting a timelapse of your newly designed holiday ornament on your Facebook, sharing your use of the technology online expands on the impact of in person outreach.

Bringing awareness to the use of the technology for new and interesting things raises the profile of the technology. It helps public perception and advance the understanding of manufacturing as a whole.

Need some more inspiration on what to post and how? Here are just a few of my favorite 3D printing creators to check out:

Estefanie Explains at All: A channel covering many aspects of making, Estefanie creates fun videos showing the process of building some crazy projects in her own kitchen. Check out her Daft Punk Helmet video for a true depiction of using 3D printing for a cool costume creation. https://www.estefannie.com/

3D Printing Nerd: Primarily on YouTube, Joel Telling makes content for 3D printing enthusiasts, and those new to 3D printing. He does videos about topics like testing new printers, trying out new 3D printing models, and looking at the latest innovations in the industry. http://the3dprintingnerd.com/

Tested: Featuring Adam Savage of Mythbusters fame, this channel covers a lot more than just 3D printing, but is the perfect example of how to document the process of making something. And they have great 3D printing content as well. https://www.tested.com/

Amie DD: Ready for some amazing 3D printed costumes? Be sure to check out Amie on social media. As a bonus, she was a contestant on the LEGO Masters reality TV show! https://www.amiedd.com/

There are so many more, but these creators are great examples.

No matter what you are 3D printing or designing, join me on Instagram, TikTok, Twitter (@erinwinick), or your favorite site in telling everyone about it. What you print will inspire others and show the world the power of 3D printing.

References

1. Winick, E.: These Amazing Met Gala Looks Took More than a Thousand Hours of 3D Printing. MIT Technology Review. https://www.technologyreview.com/2019/05/07/239148/these-amazing-met-gala-looks-took-more-than-a-thousands-hours-of-3d-printing/ (2019). Accessed 15 July 2020
2. Winick, E.: Hot off the Printer: Moving 3D-Printed Fashion from the Runway to the Rack. In: Technology Shaping the Fashion Industry. Medium. https://medium.com/s/technology-shaping-the-fashion-industry/hot-off-the-printer-af45d156745e (2017). Accessed 27 June 2020

3. Winick, E.: I 3D-Printed Every Bit of my Wedding—Including my Bouquet. MIT Technology Review.   https://www.technologyreview.com/2018/11/30/138864/i-3-d-printed-every-bit-of-my-weddingincluding-my-bouquet/ (2018). Accessed 11 July 2020

Erin Winick Anthony is a maker, engineer, writer, entrepreneur, science fashionista, and science communicator. Her passion is helping people engage with science and engineering in fun new ways, and helping communicate complex science topics.

Erin served as the CEO of Sci Chic, a company she founded in 2015 which creates plastic and metal 3D printed jewelry that is inspired by science and engineering concepts. She created the company while in college as a tool to teach about science, technology, and manufacturing, and to spark conversations about science. She grew Sci Chic to becoming profitable business by the time she graduated, running the business full time. She is the winner of the 2016 national #CampusPitch Competition and the 2017 TCU Values and Ventures Elevator Pitch Competition for her work with Sci Chic. She was also chosen as one of the Society of Manufacturing Engineers' 2017 30 Under 30 in Manufacturing.

Erin now works a full-time job as a science communicator for the International Space Station, communicating the science performed aboard the orbiting laboratory. She assigns, writes, and edits stories and video scripts for publication on NASA.gov covering station science. She also handles the social media and distribution strategy for station science content, running the @ISS_Research Twitter account, and composing science content for other NASA accounts. Erin also supports astronauts with science communication materials.

As a maker, Erin has created everything from 3D printed wedding bouquets to Ms. Frizzle costumes. She features these projects, as well her science-inspired fashion, on her Instagram and Twitter (@erinwinick) to serve as a science communication tool. Her work has been featured everywhere from CNN Money to The Daily Dot to Inverse. She has published research on 3D printing outreach in partnership with the University of Florida (UF)'s Marston Library.

She previously worked as a freelance science writer, the space reporter for the MIT Technology Review, and Technology Review's associate editor of the future of work. She developed and wrote The Airlock, a weekly email on emerging space technologies for Technology Review. Erin has helmed the publication's daily tech newsletter, The Download, ran Tech Review's Instagram account, and launched Clocking In, the publication's future work newsletter. During her freelance career, she has written stories and scripts for Engineering.com, Beanz, SciShow, IEEE Potentials, Medium, and created a Computer Aided Design course for LinkedIn Learning. She previously interned as the Richard Casement Intern for The Economist's science and technology section.

Erin graduated from UF with a bachelor's degree in mechanical engineering 2016. As an engineering student, Erin interned at John Deere, Solar Turbines, Keysight Technologies, and Bracken Engineering. During her time at Solar Turbines and Keysight, she helped introduce 3D printing to various processes on the manufacturing floor.

Erin has continually served as an advocate for women in STEM. She served as the president of UF's Society of Women Engineers and the SWE Region D Collegiate Representative from 2015-2016. She has volunteered at local elementary schools, museums, and maker events, bringing accessible science to the next generation.

Erin spends her weekends hiking, writing, 3D printing, playing with her cat, listening to Broadway cast albums, and speaking at events to spread science literacy and the art of science.

Erinwinick.com/@erinwinick/SciChic.com

Correction to: Ceramic Additive for Aerospace

Lisa Rueschhoff

Correction to:
Chapter 10 in: S. M. DelVecchio (ed.), *Women in 3D Printing*,
Women in Engineering and Science,
https://doi.org/10.1007/978-3-030-70736-1_10

The author inadvertently missed to update the release of the chapter authorized by the government in the proof. The information has been updated as "Distribution A: Cleared for Public Release, #AFRL-2021-0114" in the Acknowledgements section.

The updated version of the chapter can be found at
https://doi.org/10.1007/978-3-030-70736-1_10

Index

© Springer Nature Switzerland AG 2021
S. M. DelVecchio (ed.), *Women in 3D Printing*, Women in Engineering and
Science, https://doi.org/10.1007/978-3-030-70736-1